Invertebrates

By the same authors
Animal Types 2 : Vertebrates

Animal Types 1

Invertebrates

Maureen A. Robinson and Judith F. Wiggins

HUTCHINSON

London Melbourne Sydney Auckland Johannesburg

Hutchinson & Co. (Publishers) Ltd

An imprint of the Hutchinson Publishing Group

62-65 Chandos Place, London WC2N 4NW

Hutchinson Publishing Group (Australia) Pty Ltd
16-22 Church Street, Hawthorn, Melbourne, Victoria 3122

Hutchinson Group (NZ) Ltd
32-34 View Road, PO Box 40-086, Glenfield, Auckland 10

Hutchinson Group (SA) (Pty) Ltd
PO Box 337, Bergvlei 2012, South Africa

First published 1970
Revised edition 1973
Reprinted 1975, 1977, 1979, 1980, 1982, 1983, 1984, 1985

© Maureen A. Robinson and Judith F. Wiggins, 1970

Experimental Section in revised edition 1973

Printed and bound in Great Britain by
Anchor Brendon Ltd, Tiptree, Essex

ISBN 0 09 118931 4

Contents

Contents

Contents

Preface

The main purpose of this book is to provide large, clearly labelled drawings of a wide variety of the invertebrates which are commonly used in the study of zoology.

There is a stage in the life of every zoology student when he is confronted with a vast assemblage of animals in the laboratory. Often without text he must draw, label and classify them. We believe that this book will satisfy the bewildered student's need.

In writing this book we have dealt briefly with the biology of each animal, and have drawn particular attention to the young or larval forms of each animal wherever possible. Classification has been kept to a minimum; we have shown how the animals are classified using the visible external features of anatomy.

The drawings are our own and are not intended for direct reproduction by the student, but are indicative of the standard which he could easily attain with practice. Scales have been omitted since the book is intended for use alongside actual specimens.

The book is designed for those students preparing for the G.C.E. at Advanced and Scholarship level. It will also be useful for students taking zoology and premedical courses at university.

The section 'Experimental Approach to the Invertebrates' was written by Mrs Rosemary Goodby, lecturer in Zoology at Walbrook College. This section deals with simple, yet significant, experiments which may be carried out by students in the laboratory to supplement their study of Invertebrate Types.

The authors would like to thank Mr Malcolm Jacobs for his suggestions and criticisms, the Natural History Museum and the Zoology Department at Queen Elizabeth College for their help and advice and in particular Mr Neil Bass for his information and drawings on *Nereis* larvae. Finally, we are indebted to the Science Department of Walbrook College for the use of specimens.

NOTE

The species selected from the thousands of invertebrates for description in this book are those required for the various 'A' level syllabuses. The authors regret the ommission of whole classes—and indeed phyla—but felt that the addition of a few extra species would not remedy this selectivity and the addition of many would defeat the main purpose of the book. Students who have mastered this material should seek out the general references on p.2, or the references given with each phylum.

Kingdom Animalia

Grades of organisation and evolutionary relationships within the animal kingdom

I SUBKINGDOM: PROTOZOA
Animals which consist of a single cell

PHYLUM PROTOZOA

II SUBKINGDOM: PARAZOA
Animals consisting of aggregates of cells.

PHYLUM PORIFERA

III SUBKINGDOM: METAZOA
Animals consisting of tissues and organs.

(A) GRADE: RADIATA
Animals exhibiting radial symmetry, with definite tissues but lack true mesoderm.

PHYLUM COELENTERATA

(B) GRADE: BILATERIA
Animals exhibiting bilateral symmetry, with tissues, definite organs and true mesoderm.

(i) Subgrade: Acoelomata
Animals with parenchyma (derived from mesoderm) which fills the space between gut and body wall. No vascular system.

PHYLUM PLATYHELMINTHES

(ii) Subgrade: Pseudocoelomata
Animals with a space between gut and body wall, formed by vacuolation of large mesodermal cells. The vacuoles converge and become fluid filled to form a perivisceral cavity.

PHYLUM ASCHELMINTHES

(iii) Subgrade: Coelomata
Animals with a space — a true body cavity or coelom, lined by peritoneum of mesodermal origin.

(a) SUPERPHYLUM: PROTOSTOMIA
Schizocoelomates — animals in which the true body cavity or coelom arises as a split in the mesoderm.

**PHYLA ANNELIDA
ARTHROPODA
MOLLUSCA**

(b) SUPER PHYLUM: DEUTEROSTOMIA

Enterocoelomates — animals in which a
true coelom arises from pouches enclosing
portions of archenteron.

PHYLUM ECHINODERMATA

GENERAL REFERENCES

Barnes, R.D.	1963	Invertebrate Zoology	London
Barrington, B.E.J.W.	1967	Invertebrate Structure and Function	London
Borradaile, L.A., Eastham, L.E.S. Potts, F.A. and Saunders, J.T.	1961	The Invertebrata	Cambridge
Buchsbaum, R.	1951	Animals Without Backbones	Harmondsworth and Hawes
Bullough, W.S.	1964	Practical Invertebrate Anatomy	London
Chapman, G. & Barker, W.G.	1964	Zoology	London
Guyer, M.F. & Lane, C.E.	1965	Animal Biology	New York
Larousse	1967	Encyclopaedia of Animal Life	
Vallim, J.	1966	Animal Kingdom	London
Vines, A.E. & Rees, N.	1969	Plant And Animal Biology	London
Weiss, P.B.	1967	Science Of Biology	New York

Phylum Protozoa

Free living aquatic — parasitic, saprophytic or symbiotic.
Microscopic — single cell or syncytial.
Specialised parts or organelles present.

Classes:

I FLAGELLATA (MASTIGOPHORA)

1. Free living, saprophytic rarely parasitic.
2. One type of nucleus.
3. Adult movement by flagella.
4. Method of feeding variable.

5. Pellicle present, definite shape — may be variable in some types.
6. Asexual reproduction by longitudinal binary fission.
e.g. *Euglena*

II RHIZOPODA (SARCODINA)

1. Free living rarely parasitic.

2. One type of nucleus.
3. Adult movement by pseudopodia.
4. Food capture by means of pseudopodia forming food cup.
5. Pellicle absent, variable shape.

6. Asexual reproduction by binary fission.
e.g. *Amoeba*

III CILIATA (CILIOPHORA)

1. Free living rarely parasitic.
2. Two types of nuclei.
3. Adult movement by cilia.
4. Food capture by means of cilia.
5. Pellicle present, definite shape.
6. Asexual reproduction by transverse binary fission.
7. Sexual process by conjugation.

e.g. *Paramecium*

IV SPOROZOA

1. Parasitic.
2. One type of nucleus.
3. Adult movement feeble.
4. Feeding organelles lacking.
5. Pellicle present, definite shape.
6. Asexual reproduction by sporogony and schizogony (types of binary fission) which occur within a life cycle including a sexual process.
e.g. *Monocystis*
    Plasmodium

REFERENCES FOR PHYLUM PROTOZOA

Cox, F.E.G.	1966	"Recent Advances in The Study Of Protozoa" (From *Looking At Animals Again*)	London
Mackinnon, D.L., R.S.J.	1961	Introduction To The Study of Protozoa	London
Vickerman, K. and Cox, F.E.G.	1967	The Protozoa	London

Suggested Films For Phylum Protozoa

Gaumont British Film Library.	1958	Protozoa (11 mins)	Colour
Gaumont British Instructional	1937	Paramecium (18 mins)	Black and White
Educational Foundation For Visual Aids	1969	The Role Of The Cell Membrane (18 mins) (Living Cell Series)	Colour

Euglena

. PHYLUM PROTOZOA

CLASS FLAGELLATA

(MASTIGOPHORA)

GENUS *Euglena*

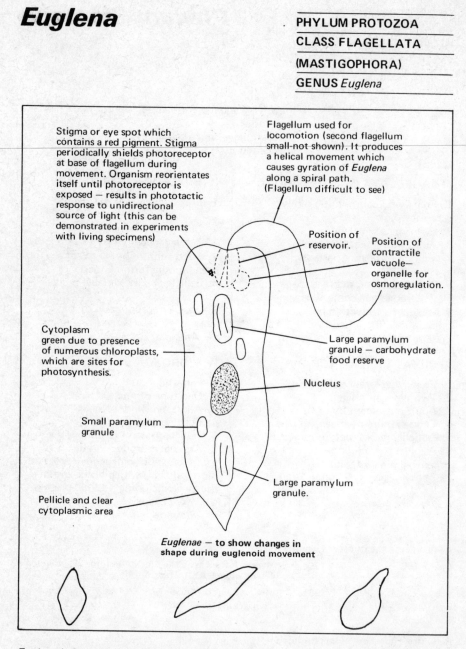

Stigma or eye spot which contains a red pigment. Stigma periodically shields photoreceptor at base of flagellum during movement. Organism reorientates itself until photoreceptor is exposed — results in phototactic response to unidirectional source of light (this can be demonstrated in experiments with living specimens)

Flagellum used for locomotion (second flagellum small-not shown). It produces a helical movement which causes gyration of *Euglena* along a spiral path. (Flagellum difficult to see)

Position of reservoir.

Position of contractile vacuole— organelle for osmoregulation.

Cytoplasm green due to presence of numerous chloroplasts, which are sites for photosynthesis.

Large paramylum granule — carbohydrate food reserve

Nucleus

Small paramylum granule

Large paramylum granule.

Pellicle and clear cytoplasmic area

Euglenae — to show changes in shape during euglenoid movement

Euglena is found in ponds and puddles. In sunlight it is able to photosynthesise its food; providing there is a ready supply of certain vitamins. In the dark however, *Euglena* is able to utilize large organic molecules for heterotrophic nutrition.

   Euglena exhibits only asexual reproduction which is by binary fission. The nucleus divides mitotically followed by longitudinal cleavage of the cytoplasm. In certain conditions *Euglena* will lose its flagella and encyst. Division may occur several times within the cyst.

Amoeba

PHYLUM PROTOZOA
CLASS RHIZOPODA
(SARCODINA)
GENUS *Amoeba*

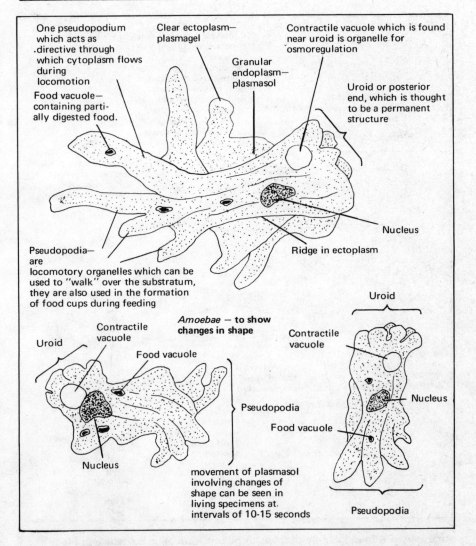

One pseudopodium which acts as .directive through which cytoplasm flows during locomotion

Food vacuole—containing partially digested food.

Clear ectoplasm—plasmagel

Granular endoplasm—plasmasol

Contractile vacuole which is found near uroid is organelle for osmoregulation

Uroid or posterior end, which is thought to be a permanent structure

Nucleus

Pseudopodia—are locomotory organelles which can be used to "walk" over the substratum, they are also used in the formation of food cups during feeding

Ridge in ectoplasm

Amoebae — to show changes in shape

Uroid

Contractile vacuole

Food vacuole

Uroid

Contractile vacuole

Nucleus

Pseudopodia

Food vacuole

Nucleus

movement of plasmasol involving changes of shape can be seen in living specimens at. intervals of 10-15 seconds

Pseudopodia

Amoeba is a free living animal (but parasitic species are known) which is found in freshwater. Although it is constantly changing its shape, there appears to be a definite anterior and posterior end. It feeds by phagocytosis or engulfing of food by the pseudopodia. The food consists of algae, and other fresh water protozoans which are enclosed in a drop of water to form a food vacuole. Asexual reproduction is by binary fission during which there is a mitotic division of the nucleus accompanied by cleavage of the cytoplasm to give two daughter amoebae. Encystment occurs in soil *Amoebae,* but not in *Amoeba proteus.*

Paramecium

PHYLUM PROTOZOA

CLASS CILIATA (CILIOPHORA)

GENUS *Paramecium*

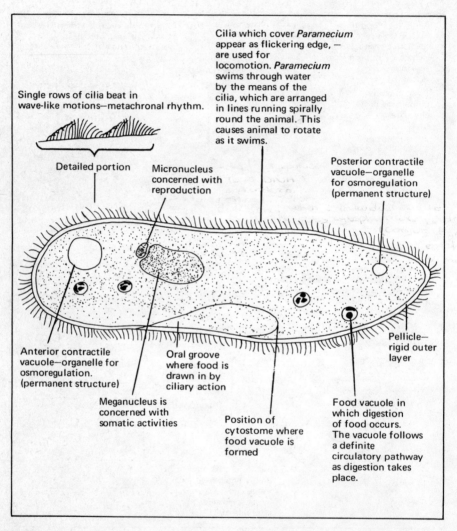

Cilia which cover *Paramecium* appear as flickering edge, — are used for locomotion. *Paramecium* swims through water by the means of the cilia, which are arranged in lines running spirally round the animal. This causes animal to rotate as it swims.

Single rows of cilia beat in wave-like motions—metachronal rhythm.

Detailed portion

Micronucleus concerned with reproduction

Posterior contractile vacuole—organelle for osmoregulation (permanent structure)

Anterior contractile vacuole—organelle for osmoregulation. (permanent structure)

Oral groove where food is drawn in by ciliary action

Pellicle— rigid outer layer

Meganucleus is concerned with somatic activities

Position of cytostome where food vacuole is formed

Food vacuole in which digestion of food occurs. The vacuole follows a definite circulatory pathway as digestion takes place.

Paramecium is found mainly in stagnant ponds, feeding on bacteria and plant particles.

Asexual reproduction is by binary fission during which the micronucleus divides mitotically, the meganucleus divides amitotically and the cytoplasm undergoes cleavage at right angles to the long axis of the body. There is a sexual process, conjugation, involving the fusion of gametic nuclei after which there is an asexual reproductive process. Another sexual process, autogamy may occur, during which gametic nuclei fuse within the same individual. In early life *Paramecium* reproduces by binary fission, then when mature by conjugation and after that by autogamy.

Monocystis

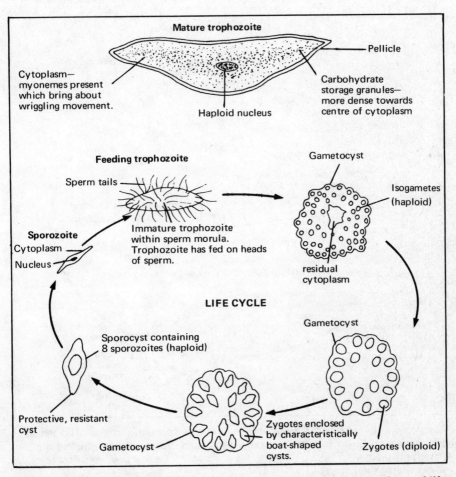

Mature trophozoite

Pellicle

Cytoplasm—
myonemes present
which bring about
wriggling movement.

Carbohydrate
storage granules—
more dense towards
centre of cytoplasm

Haploid nucleus

Feeding trophozoite

Gametocyst

Sperm tails

Isogametes
(haploid)

Sporozoite

Cytoplasm

Nucleus

Immature trophozoite
within sperm morula.
Trophozoite has fed on heads
of sperm.

residual
cytoplasm

LIFE CYCLE

Gametocyst

Sporocyst containing
8 sporozoites (haploid)

Protective, resistant
cyst

Gametocyst

Zygotes enclosed
by characteristically
boat-shaped
cysts.

Zygotes (diploid)

Monocystis is parasitic in one host only, the earthworm. It has a simple form of life
cycle since it does not require a vector. The worm becomes infected by ingesting a
sporocyst containing the sporozoites. The sporocyst is digested releasing the sporozoites which penetrate the gut wall and enter the coelom. They migrate to the
seminal vesicles where they feed within the sperm morulae and develop into mature
trophozoites.

Sexual reproduction starts by the formation of a gametocyst around two pairing
trophozoites, which divide mitotically to form isogametes. These fuse in pairs to
form diploid zygotes which undergo asexual sporogony involving meiosis to form
8 haploid sporozoites within a sporocyst. Each sporozoite will form an adult
trophozoite if eaten by another worm after the death of the original host. The
parasite appears to cause the host little inconvenience apart from reducing the
number of host's sperm.

Plasmodium

PHYLUM PROTOZOA

CLASS SPOROZOA

GENUS *Plasmodium*

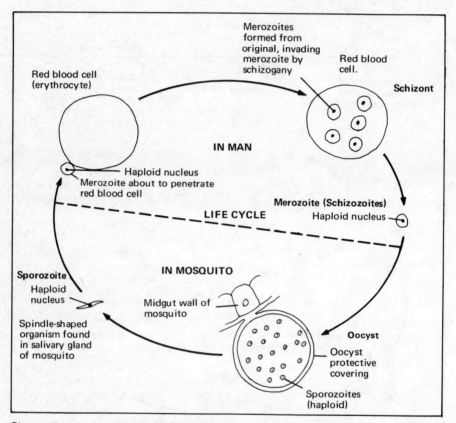

Plasmodium is the parasite responsible for the disease malaria in man. The female anopheline mosquito with biting mouth parts, is the vector which transfers the parasite from host to host by injecting it as a sporozoite in a drop of saliva prior to sucking human blood.

The sporozoite invades the human liver cell where it reproduces asexually by multiple mitotic divisions, schizogony, to form numerous haploid merozoites which are liberated by rupture of the liver cell. The merozoite then invades a red blood cell where it reproduces again asexually by schizogony (multiple mitotic divisions) to form more haploid merozoites which are released by rupture of the red blood cell and may reinvade the liver cells or penetrate more red blood cells or alternatively develop into micro- or megagametocytes within the red blood cells and undergo no further development until transferred to the stomach of a female mosquito after blood sucking. The gametocytes develop into anisogametes which fuse to form diploid, motile, zygotes which secrete a cyst. Within the cyst the zygote undergoes sporogony involving a meiotic division to form numerous haploid sporozoites which are liberated by rupture of the oocyst and migrate to the salivary glands where they remain until they are injected into human blood.

Phylum Porifera

Sponges, aquatic usually marine.
Solitary or colonial.
Most adults sedentary, with remarkable
regenerative powers.
Multicellular—consist of aggregations of cells,
not yet organised into tissues, organs or a
nervous system.
Body frequently lacks symmetry.
Variable in size, texture and colour.

External Features

Surface penetrated by many openings of
aquiferous system.
Smaller, more numerous, openings (pores) form
part of inhalant system.
Larger, fewer openings (oscula) form part of
exhalant system.

Internal Features

Part or all of aquiferous system lined by collar
cells (choanocytes).
Internal skeleton usually present, may be
composed of spongin, calcium carbonate or
silica spicules which support the body.

Asexual reproduction by budding, fragmentation
and gemmules which may form asexual larvae.
Sexual reproduction results in the formation of
a free swimming larva.

CLASS CALCAREA

Internal skeleton of calcareous spicules.
e.g. *Sycon.*

Sycon

PHYLUM PORIFERA

CLASS CALCAREA

GENUS *Sycon*

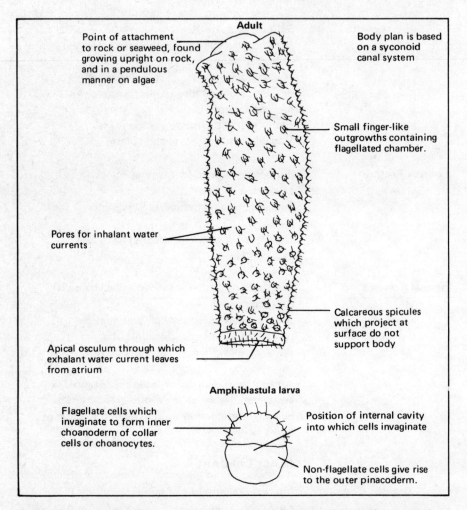

Adult

Point of attachment to rock or seaweed, found growing upright on rock, and in a pendulous manner on algae

Body plan is based on a syconoid canal system

Small finger-like outgrowths containing flagellated chamber.

Pores for inhalant water currents

Calcareous spicules which project at surface do not support body

Apical osculum through which exhalant water current leaves from atrium

Amphiblastula larva

Flagellate cells which invaginate to form inner choanoderm of collar cells or choanocytes.

Position of internal cavity into which cells invaginate

Non-flagellate cells give rise to the outer pinacoderm.

Sycon is a marine sponge found attached to rocks or seaweed in shallow coastal waters around the British coast. The body wall is folded to produce thimble like processes which are lined by flagellate collar cells or choanocytes which produce water currents and collect plankton and fine detritus particles on which the sponge feeds.

Asexual reproduction is by budding or by the formation of gemmules which are aggregations of cells which can withstand extreme conditions.

Sponges are usually hermaphrodite and protandrous. During sexual reproduction sperm are released and carried in a water current to fertilise the ovum in situ. The zygote develops into an amphiblastula larva which is released, is an aid to dispersal, and eventually settles and forms a young sponge.

Phylum Coelenterata

Aquatic, marine and fresh water.
Polymorphism exhibited.
Sedentary polyp forms, which may be solitary or colonial.
Motile medusoid forms, which are solitary.
Diploblastic, external ectoderm separated by a layer of jelly-like mesogloea, from internal endoderm.
Basic radial symmetry.

EXTERNAL FEATURES
Sac shaped bodies with one aperture which serves as mouth and anus.
Tentacles present surrounding the mouth.
Microscopic nematocysts or thread capsules present.

INTERNAL FEATURES
Nervous system consists of a nerve net either side of the mesogloea.
Single body cavity, enteron, which may be divided by mesenteries or canals.

Classes:

I HYDROZOA	II SCYPHOZOA	III ANTHOZOA
1. Solitary or colonial.	1. Solitary.	1. Solitary or colonial.
2. Polyp and medusoid forms usually present, although either may be suppressed.	2. Free medusoid swimming form, polyp form reduced or suppressed.	2. Sedentary polyp form only, medusoid form suppressed.
3. Enteron undivided in polyp forms, but divided into four canals in medusoid forms.	3. Enteron divided to form four gastric pouches leading to numerous branched and unbranched canals in the medusoid form.	3. Enteron divided by mesenteries.
4. Ectoderm not intucked through mouth to form stomodaeum. e.g. *Hydra* *Obelia*	4. Stomodaeum absent. e.g. *Aurelia* — jelly fish	4. Stomodaeum present. e.g. *Actinia* — sea anemone

Suggested Films For Phylum Coelenterata

Gaumont British Instructional	1937	Coelenterata (11 mins)	Black and White
Gaumont British Instructional	1935	Obelia (10 mins)	Black and White

Hydra

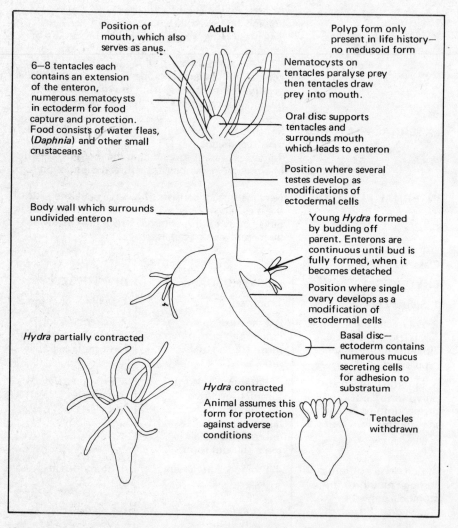

Position of mouth, which also serves as anus.

Adult

Polyp form only present in life history— no medusoid form

6–8 tentacles each contains an extension of the enteron, numerous nematocysts in ectoderm for food capture and protection. Food consists of water fleas, (*Daphnia*) and other small crustaceans

Nematocysts on tentacles paralyse prey then tentacles draw prey into mouth.

Oral disc supports tentacles and surrounds mouth which leads to enteron

Position where several testes develop as modifications of ectodermal cells

Body wall which surrounds undivided enteron

Young *Hydra* formed by budding off parent. Enterons are continuous until bud is fully formed, when it becomes detached

Position where single ovary develops as a modification of ectodermal cells

***Hydra* partially contracted**

Basal disc— ectoderm contains numerous mucus secreting cells for adhesion to substratum

***Hydra* contracted**

Animal assumes this form for protection against adverse conditions

Tentacles withdrawn

Hydra is a fresh water solitary polyp form found attached to weeds in ponds and streams. *Hydra* may be green due to presence of a symbiotic green alga *Zoochlorella*, or may be brown due to presence of a brownish alga *Zooxanthella*, in the endoderm. The gonads are temporary structures — separate sexes are found in some species of *Hydra* other species are hermaphrodite and protandrous to ensure cross-fertilisation. The zygote is retained in the ovary and develops into the embryo in situ, until it is released in a chitinous shell capable of withstanding adverse conditions. During favourable conditions the shell splits to release a young *Hydra*. Asexual reproduction by budding occurs during the warmer months of the year.

PHYLUM COELENTERATA
CLASS HYDROZOA
GENUS *Obelia*

Obelia

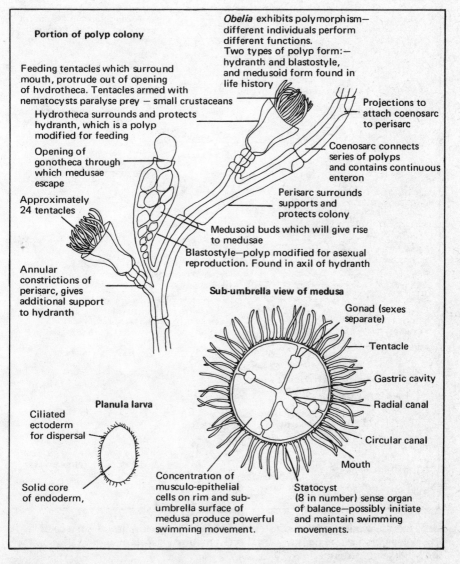

Portion of polyp colony

Obelia exhibits polymorphism—different individuals perform different functions.
Two types of polyp form:—hydranth and blastostyle, and medusoid form found in life history

Feeding tentacles which surround mouth, protrude out of opening of hydrotheca. Tentacles armed with nematocysts paralyse prey — small crustaceans

Hydrotheca surrounds and protects hydranth, which is a polyp modified for feeding

Opening of gonotheca through which medusae escape

Approximately 24 tentacles

Annular constrictions of perisarc, gives additional support to hydranth

Projections to attach coenosarc to perisarc

Coenosarc connects series of polyps and contains continuous enteron

Perisarc surrounds supports and protects colony

Medusoid buds which will give rise to medusae

Blastostyle—polyp modified for asexual reproduction. Found in axil of hydranth

Sub-umbrella view of medusa

Gonad (sexes separate)

Tentacle

Gastric cavity

Radial canal

Circular canal

Mouth

Planula larva

Ciliated ectoderm for dispersal

Solid core of endoderm,

Concentration of musculo-epithelial cells on rim and sub-umbrella surface of medusa produce powerful swimming movement.

Statocyst (8 in number) sense organ of balance—possibly initiate and maintain swimming movements.

Obelia is a sedentary marine polyp colony which produces motile, medusoid individuals. The colony is attached to sea weed in inter-tidal zone and deep water areas. Medusae are liberated into the sea and release the gametes into the water. Fertilisation results in a zygote which develops into a free swimming planula larva, which aids dispersal and then settles on sea weed and gives rise to a new polyp colony.

Aurelia
Common jellyfish

PHYLUM COELENTERATA
CLASS SCYPHOZOA
GENUS *Aurelia*

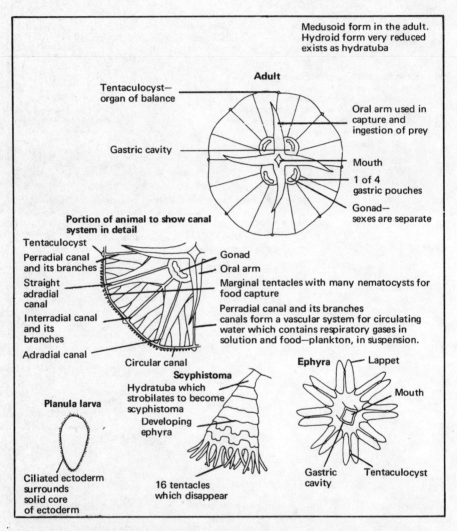

Medusoid form in the adult.
Hydroid form very reduced
exists as hydratuba

Adult

Tentaculocyst—
organ of balance

Oral arm used in
capture and
ingestion of prey

Gastric cavity

Mouth

1 of 4
gastric pouches

Gonad—
sexes are separate

**Portion of animal to show canal
system in detail**

Tentaculocyst
Perradial canal
and its branches
Straight
adradial
canal
Interradial canal
and its
branches
Adradial canal

Gonad
Oral arm
Marginal tentacles with many nematocysts for
food capture

Perradial canal and its branches
canals form a vascular system for circulating
water which contains respiratory gases in
solution and food—plankton, in suspension.

Circular canal

Ephyra — Lappet

Scyphistoma
Hydratuba which
strobilates to become
scyphistoma
Developing
ephyra

Mouth

Planula larva

Ciliated ectoderm
surrounds
solid core
of ectoderm

16 tentacles
which disappear

Gastric
cavity

Tentaculocyst

Jellyfish are marine medusoid forms found swimming close to the surface of
coastal waters feeding on plankton. Large amounts of mesogloea give the animal
its jellylike consistency.

During sexual reproduction, eggs are fertilised in the frills of the oral arms. The
zygote gives rise to a free swimming planula larva which is liberated into the sea and
settles on weeds. The planula develops into a small polyp form, the hydratuba.
During winter the hydratuba develops numerous horizontal buds and is known as a
scyphistoma. Each bud is a potential ephyra larva which is released at intervals and
gives rise to an adult medusoid form.

14

Actinia
Sea anemone

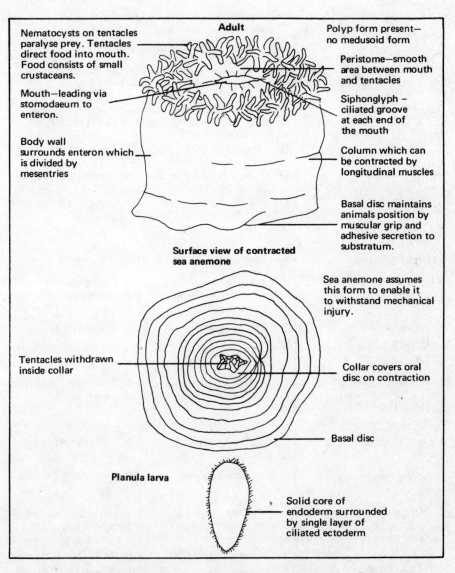

Adult

Nematocysts on tentacles paralyse prey. Tentacles direct food into mouth. Food consists of small crustaceans.

Mouth—leading via stomodaeum to enteron.

Body wall surrounds enteron which is divided by mesentries

Polyp form present— no medusoid form

Peristome—smooth area between mouth and tentacles

Siphonglyph – ciliated groove at each end of the mouth

Column which can be contracted by longitudinal muscles

Basal disc maintains animals position by muscular grip and adhesive secretion to substratum.

Surface view of contracted sea anemone

Sea anemone assumes this form to enable it to withstand mechanical injury.

Tentacles withdrawn inside collar

Collar covers oral disc on contraction

Basal disc

Planula larva

Solid core of endoderm surrounded by single layer of ciliated ectoderm

The anemone is a marine, sedentary, solitary hydroid form found between tide marks and also in deep water. Sexes are separate and fertilisation is external or within the enteron of the female. The zygote develops into a free swimming planula larva, which settles and gives rise to new polyp form. Asexual reproduction may occur by regeneration of fragments torn from the basal disc or by splitting longitudinally, starting from the mouth to form two individuals.

15

Phylum Platyhelminthes

Flat worms, free-living — aquatic or parasitic.
Triploblastic acoelomates.
Bilaterally symmetrical.

External Features

Often flattened dorsoventrally.
One aperture serving as mouth and anus.

Internal Features

Nervous system present in some forms consisting of
aggregates of nerve cells to form ganglia and two
lateral nerve cords.
Gut, if present, usually branched.
Excretory system consists of flame cells which are
linked by a network of ducts opening to exterior by
one or more pores.
Usually hermaphrodite with complicated reproductive
organs.

Classes:

I TURBELLARIA	II TREMATODA	III CESTODA
1. Free living or parasitic. Terrestrial or aquatic.	1. Parasitic, alternates between vertebrate and invertebrate hosts.	1. Endoparasitic, alternates between two animal hosts.
2. Cuticle absent.	2. Thick cuticle.	2. Thick cuticle.
3. Ciliated epidermis and rhabdites.	3. Cilia and rhabdites absent, spines often present.	3. Cilia, rhabdites and spines absent.
4. Active movement mainly by muscular action, and possibly ciliary.	4. Variable movement by muscular action.	4. Feeble movement by muscular action.
5. Mouth and gut present.	5. Mouth and gut present.	5. Mouth and gut absent.
6. Proglottids absent.	6. Proglottids absent.	6. Proglottids present.
7. Suckers rarely present.	7. Well developed oral and ventral suckers present for attachment to host.	7. Numerous suckers on scolex, and hooks present for attachment to host.
8. Indirect and direct development.	8. Several larval forms for transmission from one host to another.	8. Several larval forms for transmission from one host to another.
e.g. *Planaria*— planarian	e.g. *Fasciola*— liver fluke	e.g. *Taenia*— tapeworm

References for Phylum Platyhelminthes

Dawes, B.	1966	"Recent Researches On Fasciola hepatica" (From Looking At Animals Again)	London
Hyman, L.H.	1951	Platyhelminthes and Rhynocoela	New York

Suggested Films for Phylum Platyhelminthes

Gaumont British Instructional	1938	Fasciola (20 mins)	Black and White
Rank Film Library	1965	Flatworms (11 mins)	Colour
Rank Film Library	1965	Parasitism (11 mins)	Colour

Planaria
Planarian

PHYLUM PLATYHELMINTHES
CLASS TURBELLARIA
GENUS *Planaria*

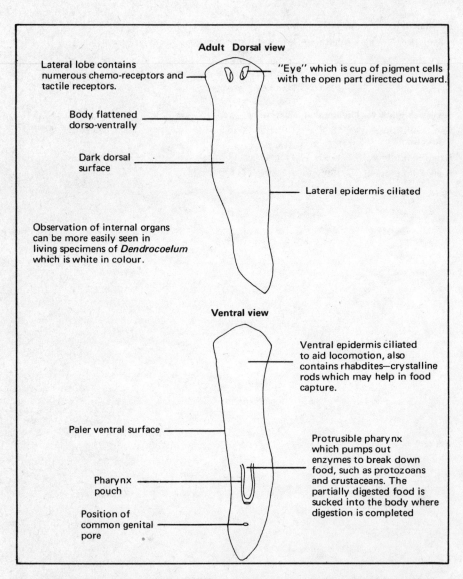

Adult   Dorsal view

Lateral lobe contains numerous chemo-receptors and tactile receptors.

"Eye" which is cup of pigment cells with the open part directed outward.

Body flattened dorso-ventrally

Dark dorsal surface

Lateral epidermis ciliated

Observation of internal organs can be more easily seen in living specimens of *Dendrocoelum* which is white in colour.

Ventral view

Ventral epidermis ciliated to aid locomotion, also contains rhabdites—crystalline rods which may help in food capture.

Paler ventral surface

Protrusible pharynx which pumps out enzymes to break down food, such as protozoans and crustaceans. The partially digested food is sucked into the body where digestion is completed

Pharynx pouch

Position of common genital pore

Planarians are free living animals found in fresh water streams, ponds and under stones. They are hermaphrodite, but pairing occurs for sperm exchange to ensure cross fertilisation. The fertilised eggs and separate yolk cells are enclosed together in a shell and give rise to miniature adults. *Planaria* also reproduces asexually either by budding or by autotomy (tearing itself into two parts) each half regenerates the missing parts.

Fasciola
Liver fluke

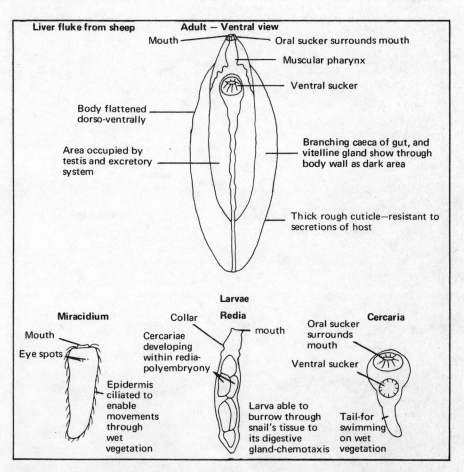

Liver fluke from sheep

Adult — Ventral view

Mouth

Oral sucker surrounds mouth

Muscular pharynx

Ventral sucker

Body flattened dorso-ventrally

Area occupied by testis and excretory system

Branching caeca of gut, and vitelline gland show through body wall as dark area

Thick rough cuticle—resistant to secretions of host

Larvae

Miracidium

Mouth

Eye spots

Epidermis ciliated to enable movements through wet vegetation

Collar

Redia

mouth

Cercariae developing within redia-polyembryony

Larva able to burrow through snail's tissue to its digestive gland-chemotaxis

Cercaria

Oral sucker surrounds mouth

Ventral sucker

Tail-for swimming on wet vegetation

Liver fluke is an endoparasite in the bile duct of sheep and less commonly in man. Sexual reproduction occurs in the vertebrate primary host. Fertilised eggs develop into miracidia within the egg capsules in the uterus. The egg capsules leave the primary host by way of the faeces and hatch to release the free-swimming miracidia which penetrate the epidermis of the invertebrate secondary host, the pond snail (*Limnaea truncatula*). In the snail's tissue the miracidium loses its cilia, assumes a spherical shape and is known as a sporocyst. Rediae are formed within the sporocyst and are liberated within the snail. They give rise to either a second generation of redia (in summer) or cercaria larvae (in winter) which leave the snail. The cercariae encyst and no further development occurs unless the cyst is swallowed by a sheep. In the small intestine of the sheep the juvenile fluke excysts by its own means and migrates to the liver by way of the coelom. It feeds on the liver cells for a while and then moves on to the bile duct where it feeds on the epithelial lining and completes its development.

19

Taenia
Tapeworm

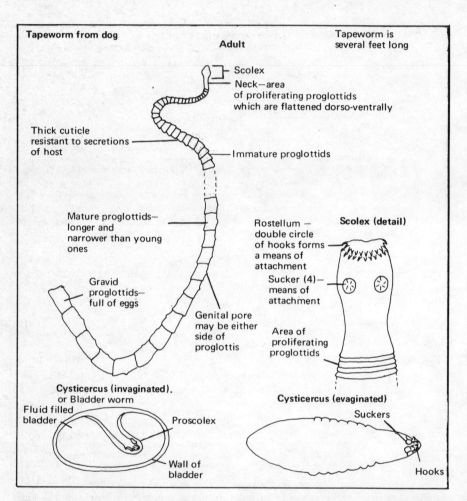

The adult tapeworm is found in the intestine of dog, the primary host. Self fertilisation results in mature proglottides containing thousands of eggs. The eggs receive a yolk cell and a thin egg shell. The egg cell divides and develops into a 6 hooked hexacanth embryo — may be ssen in T.S. of gravid proglottides. The embryo is covered by a hard embryonic membrane and is known as an onchosphere. Ripe proglottides leave the host in the faeces. The embryos do not develop any further unless they are eaten by a suitable secondary host (a rabbit or hare). In the intestine the hexacanth embryo is released from the onchosphere and bores its way through the gut wall by means of its 6 hooks, then it enters the blood stream and eventually reaches muscular tissue where it develops into the cysticercus or bladder worm. Dogs are infected by eating infected rabbit or hare. In the dog's intestine the cysticercus evaginates, attaches itself to the wall and proliferates proglottides.

Phylum Aschelminthes

Ubiquitous distribution.
Free living and parasitic forms in plants and animals.
Triploblastic psuedocoelomates—body spaces but not a true coelom present.
Bilaterally symmetrical.

CLASS NEMATODA

Roundworms, with elongated, non segmented bodies.
Absence of cilia.

External Features

Free living forms microscopic, parasitic forms tend to be larger.
Uniformly cylindrical bodies, tapering and pointed at both ends.
Cuticle of thick elastic protein.
Sense organs very primitive.
Mouth and anus present.

Internal Features

Nervous system of circumpharyngeal ring and a number of longitudinal nerve cords.
Straight unbranched gut, passing from mouth to anus.
Excretory system present, much diversity of form.
Sexes usually separate.
e.g. *Ascaris* — round worm.

Ascaris
Roundworm

PHYLUM ASCHELMINTHES

CLASS NEMATODA

GENUS *Ascaris*

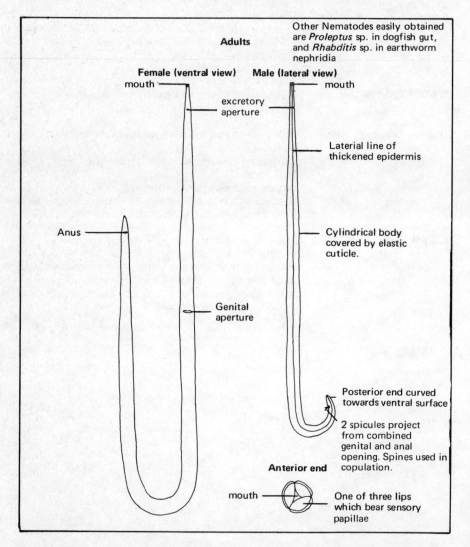

Adults

Other Nematodes easily obtained are *Proleptus* sp. in dogfish gut, and *Rhabditis* sp. in earthworm nephridia

Female (ventral view)   **Male (lateral view)**

mouth

mouth

excretory aperture

Laterial line of thickened epidermis

Anus

Cylindrical body covered by elastic cuticle.

Genital aperture

Posterior end curved towards ventral surface

2 spicules project from combined genital and anal opening. Spines used in copulation.

Anterior end

mouth

One of three lips which bear sensory papillae

The round worm is parasitic in the small intestine of mammals. It is thought to feed by sucking the host's digested food products by means of its suctorial pharynx. The sexes are usually separate and the female is longer than the male. Copulation results in fertilized eggs which pass out of the host in the faeces. On reaching a suitable host, the larvae hatch out and penetrate the epithelial lining of the small intestine. They are transported in the blood to the lungs which they penetrate causing inflammation and are later coughed up, swallowed and eventually reach the small intestine where they become adults.

Phylum Annelida

Segmented worms — aquatic and terrestrial.
Free living or parasitic.
Triploblastic coelomates.
Bilaterally symmetrical, metamerically segmented.

External Features

Body divided into a definite number of segments.
Cuticle of non-chitinous material.
Chaetae or chitinous bristles arranged segmentally.
Sense organs concentrated at anterior end.

Internal Features

Nervous system consists of solid, ventral, double nerve
cord connected by commissures to paired dorsal
cerebral ganglia.
Closed vascular system.
Respiration over whole, moist, body surface and
parapodia and gills where present.
Nephridia are usual excretory organs.

Classes:

I POLYCHAETA	II OLIGOCHAETA	III HIRUDINEA
1. Marine and estuarine — some freshwater	1. Fresh water and terrestrial.	1. Marine, fresh water and terrestrial.
2. Outer annuli correspond to inner septa.	2. Outer annuli correspond to inner septa.	2. Outer annuli more numerous than inner septa.
3. Distinct head with appendages.	3. Indistinct head without appendages.	3. Indistinct head without appendages.
4. Parapodia—projections of the body wall bearing numerous chaetae	4. Parapodia absent, fewer chaetae present arising from ectodermal sac.	4. Parapodia absent, chaetae usually absent.
5. Suckers absent.	5. Suckers absent.	5. Anterior sucker with three jaws and posterior sucker—ectoparasitic.
6. Sexes separate.	6. Hermaphrodite, but cross fertilisation occurs.	6. Hermaphrodite, but cross fertilisation occurs.
7. External fertilisation, eggs develop into trochosphere larvae. e.g. *Nereis*— ragworm *Arenicola*— lugworm	7. Eggs laid in cocoons, direct development— no larva stage. e.g. *Lumbricus*— earthworm	7. Eggs laid in cocoons, direct development— no larva stage. e.g. *Hirudo*— leech

References for Phylum Annelida

Dales, R.P.   1967   Annelids   London

Suggested Films for Phylum Annelida

Gaumont British   1935   Annelid Worms (10 mins)   Black and White
Instructional

Nereis
Ragworm

PHYLUM ANNELIDA

CLASS POLYCHAETA

GENUS *Nereis*

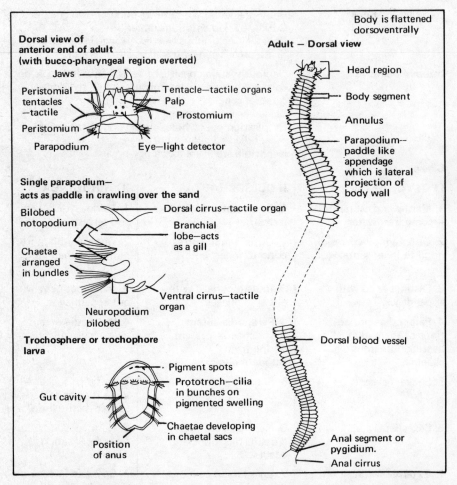

Dorsal view of anterior end of adult (with bucco-pharyngeal region everted)

Jaws

Peristomial tentacles —tactile

Tentacle—tactile organs

Palp

Prostomium

Peristomium

Parapodium

Eye—light detector

Single parapodium— acts as paddle in crawling over the sand

Bilobed notopodium

Dorsal cirrus—tactile organ

Branchial lobe—acts as a gill

Chaetae arranged in bundles

Ventral cirrus—tactile organ

Neuropodium bilobed

Trochosphere or trochophore larva

Pigment spots

Prototroch—cilia in bunches on pigmented swelling

Gut cavity

Chaetae developing in chaetal sacs

Position of anus

Body is flattened dorsoventrally

Adult — Dorsal view

Head region

Body segment

Annulus

Parapodium— paddle like appendage which is lateral projection of body wall

Dorsal blood vessel

Anal segment or pygidium.

Anal cirrus

The ragworm is a marine worm which lives in a burrow in mud or muddy sand, through which it can maintain a current of water. It feeds by making a conical net of mucus at the mouth of the burrow in which detritus and plankton become trapped, the worm eats the net containing the food. Other species of *Nereis* are predators using their jaws. In the ragworm the sexes are separate. Gametes are shed into the sea water by rupture of the body wall. After fertilisation the zygotes develop into trochospheres which remain on the surface layer of the mud and eventually give rise to adult worms.

Some species of ragworm have a special, active sexual form, the *Heteronereis* which bears enlarged eyes and enlarged parapodia on posterior segments. These form swarms in the surface waters where they liberate gametes by rupture of the body wall. These active forms make fertilisation more likely.

Arenicola
Lugworm

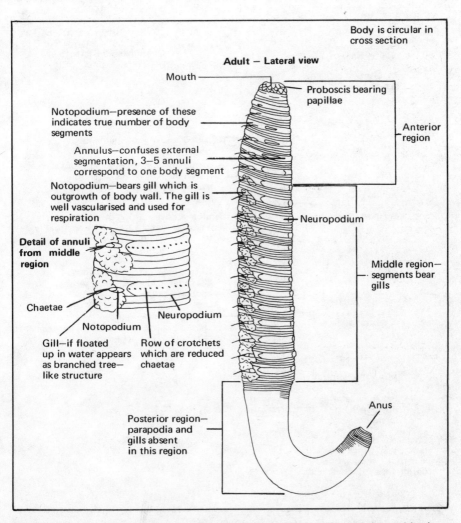

Body is circular in cross section

Adult — Lateral view

Mouth

Proboscis bearing papillae

Notopodium—presence of these indicates true number of body segments

Anterior region

Annulus—confuses external segmentation, 3—5 annuli correspond to one body segment

Notopodium—bears gill which is outgrowth of body wall. The gill is well vascularised and used for respiration

Neuropodium

Detail of annuli from middle region

Middle region— segments bear gills

Chaetae

Neuropodium

Notopodium

Gill—if floated up in water appears as branched tree— like structure

Row of crotchets which are reduced chaetae

Anus

Posterior region— parapodia and gills absent in this region

The lugworm lives in a J-shaped burrow which it makes in soft mud or sand in the inter-tidal zone and below the low tide mark. The worm cast on the surface of the sand or mud marks the end of the vertical shaft. The lugworm swallows large quantities of sand from which it digests any organic constituents. The animal replenishes the oxygen in the burrow by moving backwards up the vertical shaft of the burrow and flushing fresh water into it. Sexes are separate and gametes are released by rupture of the body wall into the burrow, and are pumped up to the surface water where fertilisation occurs. The zygote gives rise to a modified trochosphere larva which remains in the surface layer of the sand where it develops into a young worm.

Lumbricus
Earthworm

PHYLUM ANNELIDA

CLASS OLIGOCHAETA

GENUS *Lumbricus*

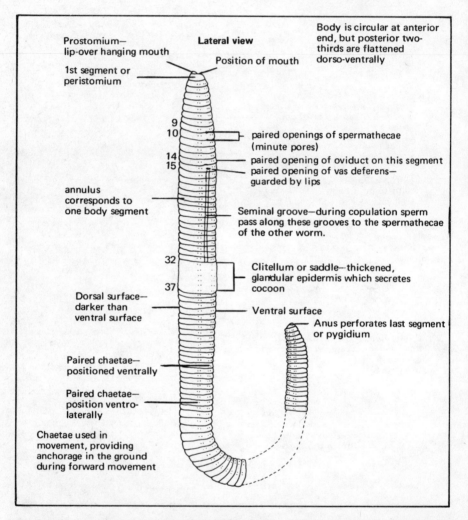

Prostomium—lip-over hanging mouth

1st segment or peristomium

Lateral view

Position of mouth

Body is circular at anterior end, but posterior two-thirds are flattened dorso-ventrally

9
10
— paired openings of spermathecae (minute pores)

14
15
— paired opening of oviduct on this segment
— paired opening of vas deferens—guarded by lips

annulus corresponds to one body segment

Seminal groove—during copulation sperm pass along these grooves to the spermathecae of the other worm.

32
Clitellum or saddle—thickened, glandular epidermis which secretes cocoon

37
Dorsal surface—darker than ventral surface

Ventral surface

Anus perforates last segment or pygidium

Paired chaetae—positioned ventrally

Paired chaetae—position ventro-laterally

Chaetae used in movement, providing anchorage in the ground during forward movement

Earthworms are terrestrial animals which burrow in soil. They swallow soil which is rich in organic material and it is on this that they feed together with vegetable matter which they collect from outside their burrow at night.

Although hermaphrodite, copulation occurs at night during which spermatozoa are exchanged between two individuals. The pairing worms separate soon after the exchange of seminal fluid has been completed.

Eggs are laid in cocoons which are dark-brown ovoid structures which are found in soil. Development is direct. There is no larval stage and the young hatch out as miniature worms. Only one or two worms develop from a single cocoon although it may originally have contained several fertilised eggs.

PHYLUM ANNELIDA

CLASS HIRUDINEA

GENUS *Hirudo*

Hirudo
Leech

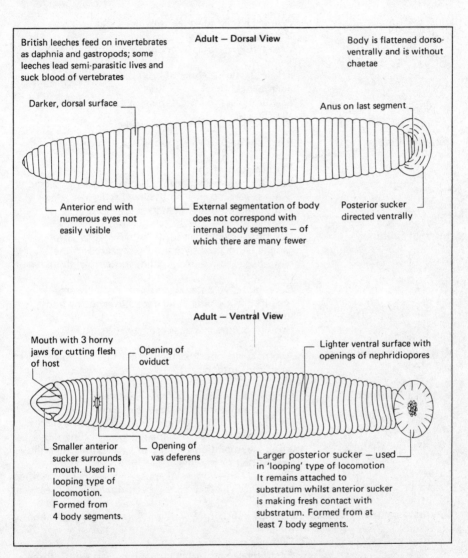

Adult — Dorsal View

British leeches feed on invertebrates as daphnia and gastropods; some leeches lead semi-parasitic lives and suck blood of vertebrates

Body is flattened dorso-ventrally and is without chaetae

Darker, dorsal surface

Anus on last segment

Anterior end with numerous eyes not easily visible

External segmentation of body does not correspond with internal body segments — of which there are many fewer

Posterior sucker directed ventrally

Adult — Ventral View

Mouth with 3 horny jaws for cutting flesh of host

Opening of oviduct

Lighter ventral surface with openings of nephridiopores

Smaller anterior sucker surrounds mouth. Used in looping type of locomotion. Formed from 4 body segments.

Opening of vas deferens

Larger posterior sucker — used in 'looping' type of locomotion It remains attached to substratum whilst anterior sucker is making fresh contact with substratum. Formed from at least 7 body segments.

Leeches are found mainly in marshes, ponds, and streams. The leech swims by vertical undulations of the body or moves over hard surfaces by looping.

It feeds on blood which it sucks from animals whilst injecting an anti-coagulant into the wound. One meal of blood can nourish the leech for many months. Although hermaphrodite, copulation occurs and sperm from one individual are transferred to the vagina of the other by a muscular penis. The eggs are laid in cocoons secreted by the clitellum. The young leeches emerge from the cocoon which may be buried in mud or attached to vegetation.

27

Phylum Arthropoda

Jointed legged invertebrates.
Ubiquitous.
Free living, parasitic, sedentary, solitary and social forms.
Triploblastic coelomates—but main body cavity a haemocoel.
Bilaterally symmetrical, metamerically segmented.
Absence of cilia.

External Features

Basically body divided into head, thorax and abdomen.
Exoskeleton of chitin.
Basically one pair of jointed appendages per segment.
Sense organs usually on the head region.
Anus situated in the posterior segment.

Internal Features

Nervous system consists of two dorsal cerebral ganglia connected by commissures around the oesophagus to a double, solid, ventral ganglionated nerve cord.
Open blood vascular system and dorsal heart.
Respiration by gills, lung books, trachea or body surface.
Growth occurs in stages after ecdysis.

I CLASS CRUSTACEA

Aquatic arthropods with two pairs of antennae, (first pair tiny—antennules).

External Features

Cephalothorax formed from head region fused to a variable number of thoracic segments.
Strong exoskeleton divided into articulating plates.
Head bears five pairs of appendages—antennules, antennae, mandibles, first pair of maxillae (maxillules), second pair of maxillae.
Thoracic and abdominal appendages may be modified as jaws, legs or accessory reproductive structures.

Internal Features

Respiration by means of gills or through general body surface.
Sexes usually separate.
First larval stage usually a nauplius larva with three pairs of appendages.

Subclasses:

BRANCHIOPODA	CIRRIPEDIA	MALACOSTRACA
1. Free living, fresh water .	1. Sedentary, marine.	1. Free living mainly marine.
2. Single compound sessile eye.	2. Adult usually without eyes.	2. Typically one pair of stalked compound eyes.
3. Carapace consists of two valves which enclose the animal and open mid-ventrally.	3. Carapace with calcareous plates embedded, encloses the trunk.	3. Carapace covers the thorax.
4. At least four pairs of thoracic appendages, fringed with bristles for filter feeding.	4. Six pairs of thoracic appendages, fringed with bristles for filter feeding —cirri.	4. Eight pairs of thoracic appendages adapted for walking or feeding.
5. Abdomen and abdominal appendages absent.	5. Abdomen and abdominal appendages absent.	5. Abdomen present, bearing appendages for swimming.
6. Sexes separate. e.g. *Daphnia—* water flea	6. Hermaphrodite. e.g. *Lepas—* goose barnacle *Balanus—* acorn barnacle	6. Sexes separate. e.g. *Astacus—* crayfish *Palaemon (Leander) —* prawn *Carcinus—* crab *Oniscus—* wood louse

References for Class Crustacea

Barrett, J, and Yonge, C.M.	1965	Collins Pocket Guide To The Sea Shore	London
Green, J.	1967	Biology of Crustacea	London
Macan, T.T.	1964	Fresh Water Invertebrate Animals	London
Street, P.	1966	Crab and Its Relatives	London

Suggested Films For Class Crustacea

Encyclopaedia Britannica Films	1960	Crustaceans (13 mins)	Colour
Rank Film Library	1967	Jointed Legged Animals (10 mins)	Colour
Rank Audio-Visual	1968	Life Story of Water Flea (Daphnia)	Colour

Daphnia
Water flea

PHYLUM ARTHROPODA

CLASS CRUSTACEA

SUBCLASS BRANCHIOPODA

GENUS *Daphnia*

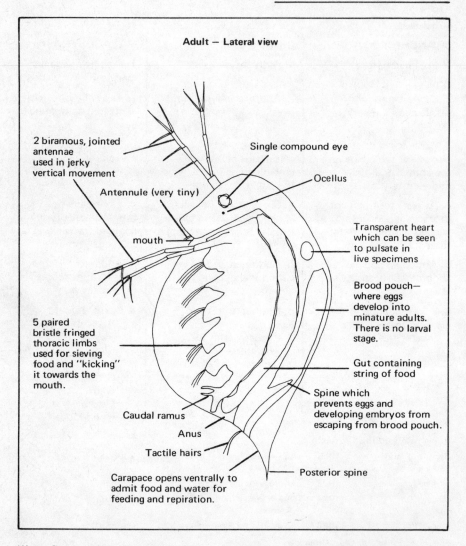

Adult — Lateral view

2 biramous, jointed antennae used in jerky vertical movement

Antennule (very tiny)

mouth

5 paired bristle fringed thoracic limbs used for sieving food and "kicking" it towards the mouth.

Single compound eye

Ocellus

Transparent heart which can be seen to pulsate in live specimens

Brood pouch— where eggs develop into minature adults. There is no larval stage.

Gut containing string of food

Spine which prevents eggs and developing embryos from escaping from brood pouch.

Caudal ramus

Anus

Tactile hairs

Carapace opens ventrally to admit food and water for feeding and repiration.

Posterior spine

Water fleas are found on the surface water of ponds and lakes. They feed on microscopic plants and particles of detritus.

Males are uncommon in this genus and parthenogenetic eggs hatch into females for several generations until certain adverse conditions induce the appearance of males in the population. As a result, a few large fertilised eggs are produced. These eggs have a thick resistant shell and much yolk and are able to withstand drought and freezing. When suitable conditions return, the eggs hatch out as female water fleas which reproduce by parthenogenesis.

Balanus
Acorn barnacle

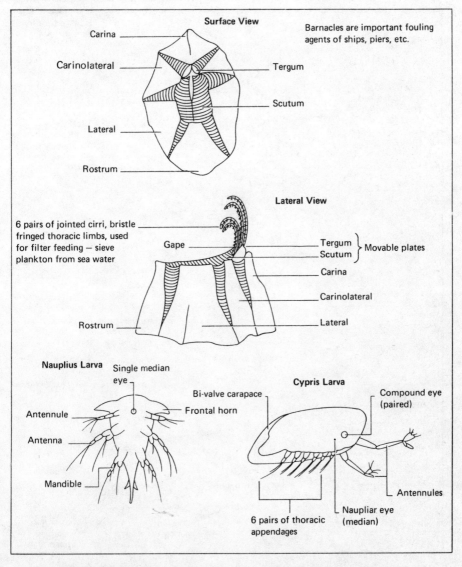

Surface View

Barnacles are important fouling agents of ships, piers, etc.

Carina

Carinolateral

Tergum

Scutum

Lateral

Rostrum

Lateral View

6 pairs of jointed cirri, bristle fringed thoracic limbs, used for filter feeding — sieve plankton from sea water

Gape

Tergum
Scutum } Movable plates

Carina

Carinolateral

Lateral

Rostrum

Nauplius Larva

Single median eye

Antennule

Antenna

Frontal horn

Mandible

Cypris Larva

Bi-valve carapace

Compound eye (paired)

6 pairs of thoracic appendages

Naupliar eye (median)

Antennules

Acorn barnacles are found covering rocks and shells at various sea depths. They can tolerate temporary exposure at low tide by closing the gape.

Although hermaphrodite, mating occurs to give fertilised eggs which hatch as free swimming nauplius larvae which moult to form cypris larvae. The cypris larvae select a suitable substratum, to which they attach themselves by using the cement glands in the antennules. Then metamorphosis into the adult form takes place.

31

Lepas
Goose barnacle

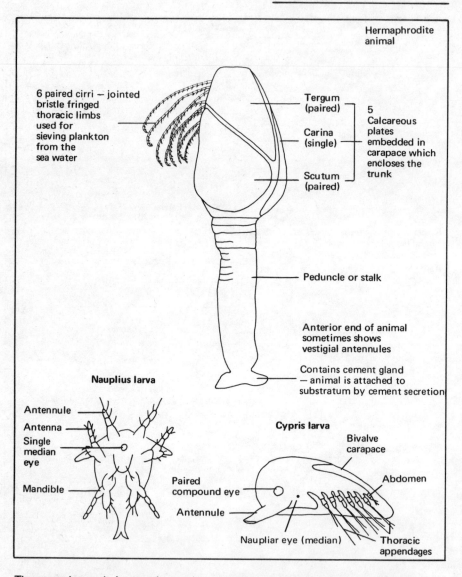

Hermaphrodite animal

6 paired cirri — jointed bristle fringed thoracic limbs used for sieving plankton from the sea water

Tergum (paired)

Carina (single)

Scutum (paired)

5 Calcareous plates embedded in carapace which encloses the trunk

Peduncle or stalk

Anterior end of animal sometimes shows vestigial antennules

Contains cement gland — animal is attached to substratum by cement secretion

Nauplius larva

Antennule

Antenna

Single median eye

Mandible

Cypris larva

Bivalve carapace

Abdomen

Paired compound eye

Antennule

Naupliar eye (median)

Thoracic appendages

The goose barnacle is a marine, sedentary animal which is found attached to floating objects, such as bottles, wood and ships.

Although the animals are hermaphrodite, mating resulting in cross fertilisation usually occurs. The young hatch out as nauplius larvae, which moult several times to form cypris larvae. The cypris larvae seek out suitable substrata and metamorphose into the adults.

Astacus
Crayfish

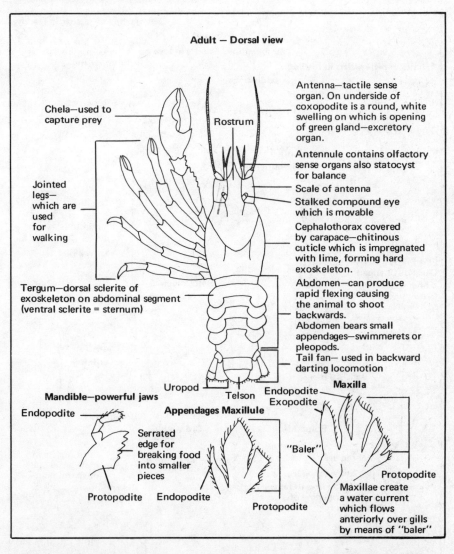

Adult — Dorsal view

Chela—used to capture prey

Rostrum

Antenna—tactile sense organ. On underside of coxopodite is a round, white swelling on which is opening of green gland—excretory organ.

Antennule contains olfactory sense organs also statocyst for balance

Scale of antenna

Stalked compound eye which is movable

Cephalothorax covered by carapace—chitinous cuticle which is impregnated with lime, forming hard exoskeleton.

Jointed legs— which are used for walking

Tergum—dorsal sclerite of exoskeleton on abdominal segment (ventral sclerite = sternum)

Abdomen—can produce rapid flexing causing the animal to shoot backwards.

Abdomen bears small appendages—swimmerets or pleopods.

Tail fan— used in backward darting locomotion

Uropod

Telson

Maxilla

Endopodite

Exopodite

Mandible—powerful jaws

Appendages Maxillule

Endopodite

Serrated edge for breaking food into smaller pieces

"Baler"

Protopodite

Protopodite

Endopodite

Protopodite

Maxillae create a water current which flows anteriorly over gills by means of "baler"

Crayfish are aquatic animals found in freshwater streams in which there is a high concentration of calcareous matter. They remain hidden by day, but emerge at night to feed on dead or living animals and plants.

The sexes are separate and fertilised eggs are attached to the female swimmerets or pleopods on the ventral surface of the abdomen where they hatch and develop into miniature adults. The female is said to be "in berry" when carrying eggs. The immature crayfish maintain their grip on the swimmerets by means of their chelae.

33

Astacus
Crayfish

PHYLUM ARTHROPODA
CLASS CRUSTACEA
SUBCLASS MALACOSTRACA
GENUS *Astacus*

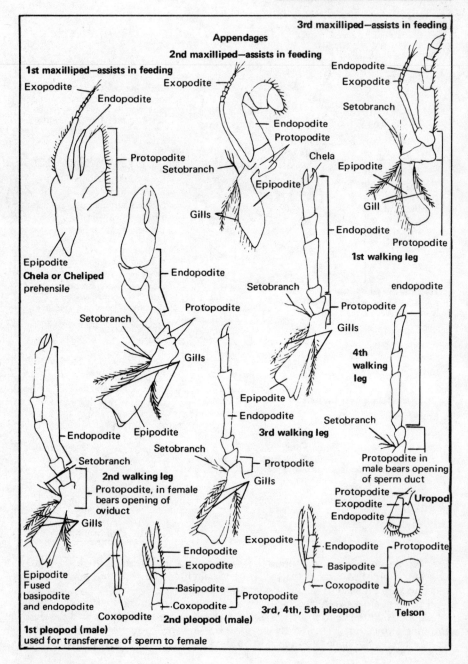

Appendages

1st maxilliped—assists in feeding
- Exopodite
- Endopodite
- Protopodite
- Setobranch
- Epipodite
- **Chela or Cheliped** prehensile

2nd maxilliped—assists in feeding
- Exopodite
- Endopodite
- Protopodite
- Setobranch
- Epipodite
- Gills

3rd maxilliped—assists in feeding
- Endopodite
- Exopodite
- Setobranch
- Epipodite
- Gill
- Endopodite
- Protopodite

1st walking leg
- Chela
- Setobranch
- Protopodite
- Gills

2nd walking leg
- Endopodite
- Protopodite
- Setobranch
- Gills
- Epipodite
- Endopodite
- Setobranch
- **Protopodite, in female bears opening of oviduct**
- Gills

3rd walking leg
- Epipodite
- Endopodite
- Protopodite
- Gills

4th walking leg
- endopodite
- Setobranch
- **Protopodite in male bears opening of sperm duct**

1st pleopod (male) used for transference of sperm to female
- Epipodite
- Fused basipodite and endopodite
- Coxopodite

2nd pleopod (male)
- Endopodite
- Exopodite
- Basipodite
- Coxopodite
- Protopodite

3rd, 4th, 5th pleopod
- Exopodite
- Endopodite
- Basipodite
- Coxopodite
- Protopodite

Uropod
- Protopodite
- Exopodite
- Endopodite
- Protopodite

Telson

PHYLUM ARTHROPODA

CLASS CRUSTACEA

SUBCLASS MALACOSTRACA

GENUS *Carcinus*

Carcinus
Shore crab

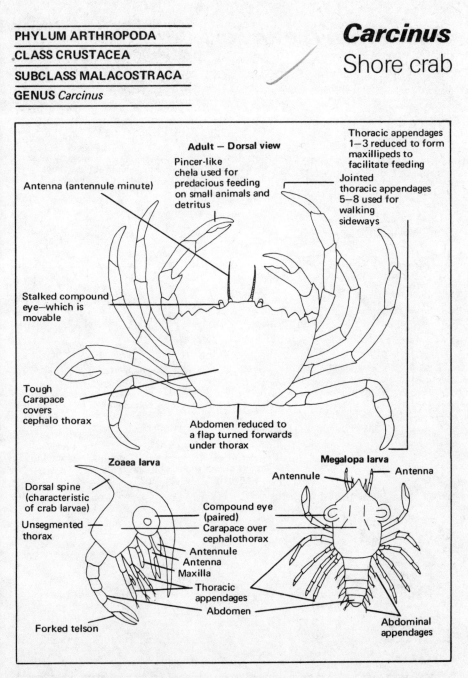

Adult — Dorsal view

Thoracic appendages 1—3 reduced to form maxillipeds to facilitate feeding

Pincer-like chela used for predacious feeding on small animals and detritus

Jointed thoracic appendages 5—8 used for walking sideways

Antenna (antennule minute)

Stalked compound eye—which is movable

Tough Carapace covers cephalo thorax

Abdomen reduced to a flap turned forwards under thorax

Zoaea larva

Dorsal spine (characteristic of crab larvae)

Unsegmented thorax

Compound eye (paired)

Carapace over cephalothorax

Antennule

Antenna

Maxilla

Thoracic appendages

Abdomen

Forked telson

Megalopa larva

Antennule

Antenna

Abdominal appendages

The shore crab is a very common inhabitant of beaches and rocky shores. Food is grasped by the chelae and passed to the third maxillipeds which push it between the other mouthparts where it is torn into manageable pieces.

Sexes are separate. Sexual reproduction results in fertilised eggs which hatch out as free swimming zoaea which develop into the second larval form, the megalopa which metamorphose into the adult.

Palaemon (Leander)
Common prawn

PHYLUM ARTHROPODA

CLASS CRUSTACEA

SUBCLASS MALACOSTRACA

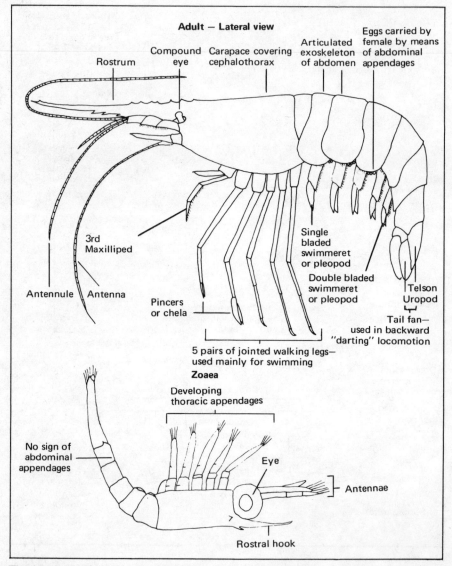

Adult — Lateral view

Rostrum

Compound eye

Carapace covering cephalothorax

Articulated exoskeleton of abdomen

Eggs carried by female by means of abdominal appendages

3rd Maxilliped

Antennule Antenna

Pincers or chela

Single bladed swimmeret or pleopod

Double bladed swimmeret or pleopod

Telson

Uropod

Tail fan— used in backward "darting" locomotion

5 pairs of jointed walking legs— used mainly for swimming

Zoaea

Developing thoracic appendages

No sign of abdominal appendages

Eye

Antennae

Rostral hook

The prawn is a marine animal found in rock pools and in the sublittoral zone. It is a scavenger and collects food such as small animals and detritus using its chelate walking legs.

Eggs hatch out as zoaea larvae which moult several times. There are several stages of zoaea, which gain increasing numbers of appendages with age until they have acquired the full adult complement.

Oniscus
Woodlouse

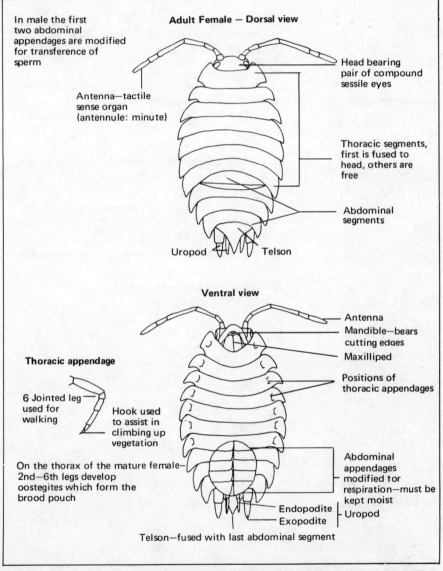

In male the first two abdominal appendages are modified for transference of sperm

Adult Female — Dorsal view

Antenna—tactile sense organ (antennule: minute)

Head bearing pair of compound sessile eyes

Thoracic segments, first is fused to head, others are free

Abdominal segments

Uropod   Telson

Ventral view

Antenna

Mandible—bears cutting edges

Maxilliped

Thoracic appendage

6 Jointed leg used for walking

Hook used to assist in climbing up vegetation

Positions of thoracic appendages

On the thorax of the mature female—2nd—6th legs develop oostegites which form the brood pouch

Abdominal appendages modified for respiration—must be kept moist

Uropod

Endopodite
Exopodite

Telson—fused with last abdominal segment

The woodlouse is a terrestrial animal found in damp places as beneath stones and dead tree trunks. It feeds on decaying vegetable or animal matter. Some woodlice have the ability to roll up into a ball, which may afford some protection against water loss or predators. Sexual reproduction results in fertilised eggs which are retained in the brood pouch until they emerge as miniature adults.

II Class Myriapoda

Usually terrestrial.

External Features

Distinct head bearing one pair of both antennae and mandibles, and one or two pairs of maxillae.
Body segments numerous, bearing one pair of appendages per segment.
Eyes if present always simple.

Internal Features

Respiration by means of trachea.
Sexes separate—young hatch as miniature adults.

Sub Class Chilopoda

1. Flattened dorsoventrally.

2. Distinct head, bearing one pair of antennae, one pair of mandibles, two pairs of maxillae and groups of simple eyes (ocelli)

3. First body segment bearing one pair of poison claws, all except the last two segments bearing one pair of walking legs.

4. Penultimate segment bearing gonopods and genital opening.

5. Last segment bearing anus.
e.g. *Lithobius* — centipede.

Lithobius
Centipede

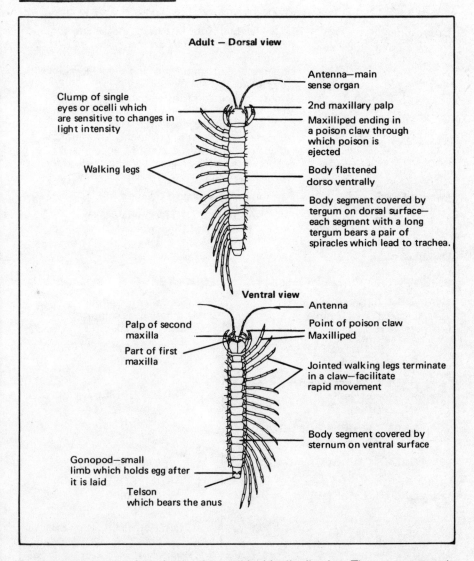

Adult — Dorsal view

Antenna—main sense organ

Clump of single eyes or ocelli which are sensitive to changes in light intensity

2nd maxillary palp

Maxilliped ending in a poison claw through which poison is ejected

Walking legs

Body flattened dorso ventrally

Body segment covered by tergum on dorsal surface— each segment with a long tergum bears a pair of spiracles which lead to trachea.

Ventral view

Antenna

Palp of second maxilla

Point of poison claw

Maxilliped

Part of first maxilla

Jointed walking legs terminate in a claw—facilitate rapid movement

Body segment covered by sternum on ventral surface

Gonopod—small limb which holds egg after it is laid

Telson which bears the anus

Centipedes are terrestrial animals with world wide distribution. They are nocturnal, hiding by day in damp places as under leaf litter, logs and stones. They are predacious and hunt insects and worms which they kill by means of poison claws.

Sexes are separate. Fertilised eggs are laid singly and buried in the earth. Development is direct and the young emerge as miniature adults, but with a reduced number of segments and only seven pairs of legs.

III Class Insecta

Flying invertebrates, with one pair of antennae.

External Features Body clearly divided into head, thorax and abdomen. Head bears paired antennae, mandibles, maxillae and labium.
One pair of compound eyes and simple eyes (ocelli) are present.
Thorax composed of three segments, bearing three pairs of walking legs and two pairs of wings on the second and third segments.
Abdomen composed of eleven segments—not all of which are demarcated externally, appendages restricted to the last segment and specialised for reproduction.

Internal Features Respiration by means of tracheae.

SUBCLASS PTERYGOTA (metabola)
Usually two pairs of wings present.

Divisions:

EXOPTERYGOTA (hemimetabola)

1. Wings develop externally.
2. Metamorphosis incomplete—egg, nymph, adult.

e.g. ORDER DICTYOPTERA
  — cockroach

ORDER ORTHOPTERA
  — locust
  — grasshopper

ORDER EPHEMEROPTERA
  — mayfly

ORDER ODONATA
  — dragonfly

ORDER HEMIPTERA
  — bug
  — aphid

ENDOPTERYGOTA (holometabola)

1. Wings develop internally.
2. Metamorphosis complete—egg, larva, pupa, adult.

e.g. ORDER LEPIDOPTERA
  — butterfly
  — moth

ORDER DIPTERA
  — housefly
  — cranefly
  — fruitfly
  — gnat
  — mosquito

ORDER COLEOPTERA
  — beetle

ORDER HYMENOPTERA
  — bee
  — ant

ORDER TRICHOPTERA
  — caddis fly

ORDER SIPHONAPTERA
  — flea

References for Class Insecta

Barrass, R.	1968	The Locust	London
Butler, E.A.	1923	Biology of the British Hemiptera and Heteroptera	London
Edwards, Oldroyd and Smart	1936	British Blood-Sucking Flies	Trustees of The British Museum
Farb, P.	1964	The Insects	Netherlands
Haskell, G.	1961	Practical Heredity with Drosophila	London
Imms, A.D.	1961	Outlines of Entomology	London
Marshall, J.F.	1939	British Mosquitoes	Trustees of The British Museum
Smart,	1956	Insects Of Medical Importance	Trustees of The British Museum
Southward, T.R.E. and Leston, D.	1959	Land And Water Bugs Of The British Isles	London

Suggested Films For Class Insecta

Gateway	1964	Aphids and Ladybirds (14 mins)	Colour
National Committee For Visual Aids In Education	1950	Gnats (5 mins)	Black and White
Gaumont British Film Library	1968	Honey Bee (10 mins)	Black and White
Rank Film Library	1960	House Fly (17 mins)	Black and White
Encyclopaedia Britannica	1966	Life Story of the Moth (The Silkworm) (11 mins)	Colour
Petroleum Films Bureau	1956	The Ruthless One (Locusts) (18 mins)	Colour

DIVISION EXOPTERYGOTA (hemimetabola)

I ORDER DICTYOPTERA

1. Forewings thickened to form tegmina, hind pair of wings membranous, covered by tegmina.
2. Mouth parts specialised for biting.
3. Incomplete metamorphosis, eggs carried by female in ootheca, nymphs terrestrial.
4. Antennae long.
5. Cerci long, many segmented.
6. Legs similar, modified for running.

e.g. *Periplaneta* — cockroach

II ORDER ORTHOPTERA

1. Forewings thickened to form tegmina which overlap each other when folded, hind wings membranous.
2. Mouth parts specialised for biting.
3. Incomplete metamorphosis, nymphs terrestrial.

4. Antennae long.
5. Cerci short, usually unsegmented.
6. Third pair of legs modified for jumping, often with stridulatory organs.
e.g. *Locusta* — locust
Chorthippus — grasshopper

III ORDER EPHEMEROPTERA

1. Two pairs of membranous wings, hind wings small or atrophied.

2. Mouthparts atrophied.

3. Incomplete metamorphosis, nymphs aquatic with tracheal gills on the abdomen.

4. Antennae minute.
5. Cerci slender, very long, many segmented, usually accompanied by a median caudal filament.
6. Legs only used whilst clinging to objects whilst resting.
e.g. *Ephemera* — mayfly

IV ORDER ODONATA

1. Two pairs of large membranous wings of equal size, both bearing cross veins.
2. Mouthparts specialised for biting and strongly toothed.
3. Incomplete metamorphosis, nymphs aquatic with tracheal gills—rectal branchial basket, labium modified to form rectractile prehensile mask.
4. Antennae minute.
5. Cerci short, consist of one segment.

6. Legs bear spines to capture insects during flight.
e.g. *Libellula* — dragonfly

V ORDER HEMIPTERA

1. Wings variably developed, reduced venation. Forewings usually corneous, apterous forms frequent.
2. Mouthparts specialised for piercing and sucking.
3. Incomplete metamorphosis, nymphs aquatic and terrestrial.
4. Antennae long.
5. Cerci short.
6. Legs modified for swimming in aquatic forms.

e.g. *Pentatoma* — forest bug
   Aphis — aphid

Periplaneta
Cockroach

PHYLUM ARTHROPODA

CLASS INSECTA

ORDER DICTYOPTERA

GENUS *Periplaneta*

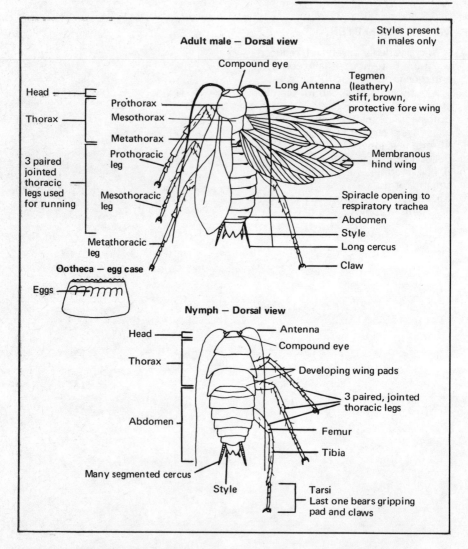

Adult male – Dorsal view

Styles present in males only

Compound eye

Long Antenna

Tegmen (leathery) stiff, brown, protective fore wing

Head

Thorax

Prothorax

Mesothorax

Metathorax

Prothoracic leg

3 paired jointed thoracic legs used for running

Mesothoracic leg

Metathoracic leg

Membranous hind wing

Spiracle opening to respiratory trachea

Abdomen

Style

Long cercus

Claw

Ootheca – egg case

Eggs

Nymph – Dorsal view

Head

Antenna

Compound eye

Developing wing pads

Thorax

3 paired, jointed thoracic legs

Abdomen

Femur

Tibia

Many segmented cercus

Style

Tarsi Last one bears gripping pad and claws

Cockroaches live in cracks and crevices of walls, particularly near boilers and cookers. They emerge at night to feed on any organic matter. They can run very rapidly, but rarely fly.

After copulation the female carries, and later deposits, a purse shaped ootheca, containing two rows of 8 eggs. Metamorphosis is incomplete and the young emerge as nymphs which resemble the adults, but they lack wings. During the next 5 years wings develop slowly from wing pads on the meso and metathorax, and the nymph moults several times until it assumes the adult form.

44

Periplaneta
Cockroach

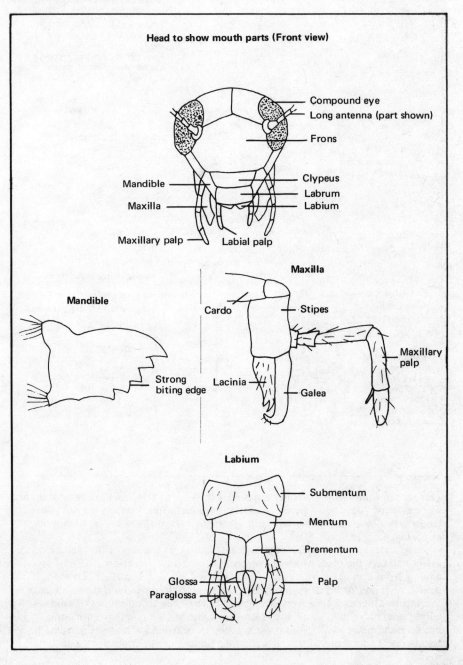

Head to show mouth parts (Front view)

Compound eye
Long antenna (part shown)
Frons
Clypeus
Mandible
Labrum
Maxilla
Labium
Maxillary palp
Labial palp

Mandible

Strong biting edge

Maxilla

Cardo
Stipes
Lacinia
Maxillary palp
Galea

Labium

Submentum
Mentum
Prementum
Glossa
Palp
Paraglossa

Locusta
Locust

PHYLUM ARTHROPODA

CLASS INSECTA

ORDER ORTHOPTERA

GENUS *Locusta*

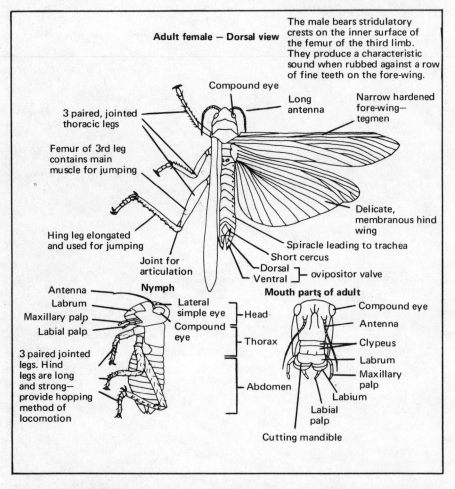

Adult female — Dorsal view

The male bears stridulatory crests on the inner surface of the femur of the third limb. They produce a characteristic sound when rubbed against a row of fine teeth on the fore-wing.

Compound eye

Long antenna

Narrow hardened fore-wing— tegmen

3 paired, jointed thoracic legs

Femur of 3rd leg contains main muscle for jumping

Hing leg elongated and used for jumping

Joint for articulation

Delicate, membranous hind wing

Spiracle leading to trachea

Short cercus

Dorsal

Ventral

ovipositor valve

Nymph

Antenna

Labrum

Maxillary palp

Labial palp

3 paired jointed legs. Hind legs are long and strong— provide hopping method of locomotion

Lateral simple eye

Compound eye

Head

Thorax

Abdomen

Mouth parts of adult

Compound eye

Antenna

Clypeus

Labrum

Maxillary palp

Labium

Labial palp

Cutting mandible

Locusts are found in the warmer regions of the world. They feed on vegetation and when swarms arise they may cause extensive devastation. Locusts are not always gregarious. Some may be solitary, but when they swarm they are of economic importance.

After mating, the female locust digs a deep hole in the sand with the ovipositor valves and lays the eggs, which are mixed with a frothy secretion. The nymphs emerge from eggs as worm like creatures and wriggle to the surface. Then they moult. Nymphs, or hoppers, moult 5 more times before reaching the adult stage.

Plagues of locusts arise from changes in environmental conditions. Good conditions, producing abundant vegetation cause large increases in the population, but sudden drought will cause overcrowding in restricted habitats. Crowding of hoppers and adults produces the gregarious behaviour which leads to a swarm.

Chorthippus
Field grasshopper

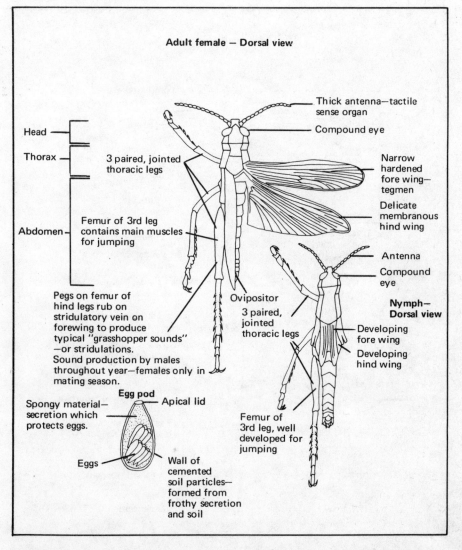

Adult female — Dorsal view

Thick antenna—tactile sense organ

Compound eye

Head

Thorax

3 paired, jointed thoracic legs

Narrow hardened fore wing— tegmen

Delicate membranous hind wing

Femur of 3rd leg contains main muscles for jumping

Abdomen

Antenna

Compound eye

Nymph— Dorsal view

Ovipositor

3 paired, jointed thoracic legs

Pegs on femur of hind legs rub on stridulatory vein on forewing to produce typical "grasshopper sounds" —or stridulations. Sound production by males throughout year—females only in mating season.

Developing fore wing

Developing hind wing

Egg pod

Spongy material— secretion which protects eggs.

Apical lid

Femur of 3rd leg, well developed for jumping

Eggs

Wall of cemented soil particles— formed from frothy secretion and soil

The field grasshopper is found in meadows and prairie lands. British grasshoppers are vegetarians and diurnal, relying on the sun's warmth for their activity.

After copulation the female lays eggs in groups of up to 14 in a tough case or egg pod, which is below the ground. The eggs are well protected to withstand the winter. A nymph enclosed in a transparent sac emerges from each egg during the spring and wriggles out of the sac then moults 4 more times before becoming a mature adult.

Ephemera
Mayfly

PHYLUM ARTHROPODA
CLASS INSECTA
ORDER EPHEMEROPTERA
GENUS *Ephemera*

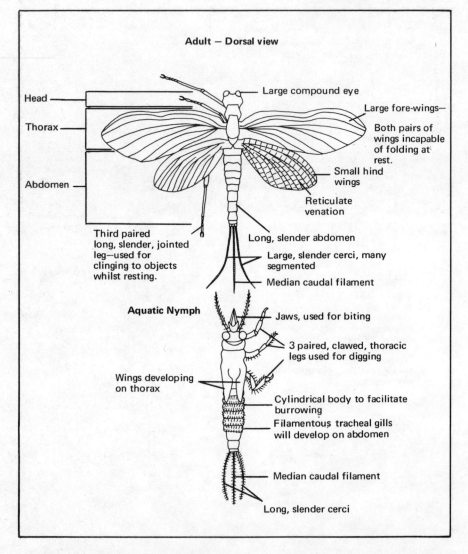

Adult — Dorsal view

Head

Thorax

Abdomen

Large compound eye

Large fore-wings—

Both pairs of wings incapable of folding at rest.

Small hind wings

Reticulate venation

Third paired long, slender, jointed leg—used for clinging to objects whilst resting.

Long, slender abdomen

Large, slender cerci, many segmented

Median caudal filament

Aquatic Nymph

Jaws, used for biting

3 paired, clawed, thoracic legs used for digging

Wings developing on thorax

Cylindrical body to facilitate burrowing

Filamentous tracheal gills will develop on abdomen

Median caudal filament

Long, slender cerci

Mayflies are found over fresh water. Eggs are laid in the water and hatch out as nymphs, which live for as long as three years on decaying matter, and burrow into the water bank. After numerous moults, the nymph emerges from the water onto a plant stem and metamorphoses into the adult.

The adult lives for a matter of hours and does not feed. (Mouth parts vestigial and alimentary canal is full of air.) During their short life mating occurs and the female lays her eggs in the water.

Libellula
Dragonfly

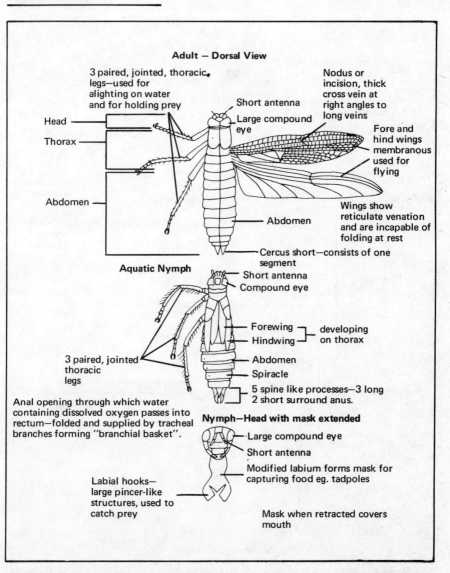

Adult — Dorsal View

3 paired, jointed, thoracic legs—used for alighting on water and for holding prey

Head

Thorax

Abdomen

Short antenna

Large compound eye

Nodus or incision, thick cross vein at right angles to long veins

Fore and hind wings membranous used for flying

Abdomen

Wings show reticulate venation and are incapable of folding at rest

Cercus short—consists of one segment

Aquatic Nymph

Short antenna

Compound eye

Forewing ⌉ developing
Hindwing ⌋ on thorax

Abdomen

Spiracle

5 spine like processes—3 long 2 short surround anus.

3 paired, jointed thoracic legs

Anal opening through which water containing dissolved oxygen passes into rectum—folded and supplied by tracheal branches forming "branchial basket".

Nymph—Head with mask extended

Large compound eye

Short antenna

Modified labium forms mask for capturing food eg. tadpoles

Labial hooks— large pincer-like structures, used to catch prey

Mask when retracted covers mouth

Dragon flies are predacious animals which feed on the wing, on other insects, such as gnats and mosquitoes. The eggs are desposited on the water and hatch out into nymphs which greatly resemble the adult. A year later the nymphs are fully grown and may be found hanging vertically downwards from a weed. The skin splits along the dorsal surface and the adult emerges. It remains on the weed whilst the wings expand and dry, and then lives for only 2—3 months.

Pentatoma
Forest bug

PHYLUM ARTHROPODA

CLASS INSECTA

ORDER HEMIPTERA

GENUS *Pentatoma*

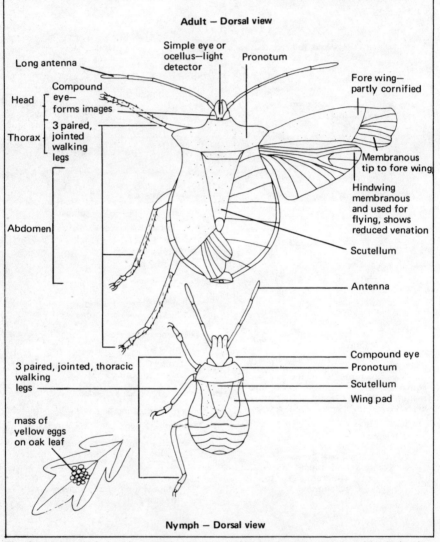

Adult — Dorsal view

Long antenna

Simple eye or ocellus—light detector

Pronotum

Head — Compound eye— forms images

Thorax — 3 paired, jointed walking legs

Abdomen

Fore wing— partly cornified

Membranous tip to fore wing

Hindwing membranous and used for flying, shows reduced venation

Scutellum

Antenna

3 paired, jointed, thoracic walking legs

Compound eye

Pronotum

Scutellum

Wing pad

mass of yellow eggs on oak leaf

Nymph — Dorsal view

The forest bug is found in woodlands and orchards; the oak and the alder being its major hosts. The mouthparts are adapted for piercing plant tissue and sucking up the sap. It also feeds on other invertebrates found on the leaf.

Spherical yellow eggs are laid in hexagonal masses of about 14 to a leaf. The eggs hatch out as nymphs which moult three times before metamorphosing into adults. Over wintering occurs before the last moult and the adults emerge in early summer.

PHYLUM ARTHROPODA

CLASS INSECTA

ORDER HEMIPTERA

GENUS *Aphis*

Aphis
Greenfly

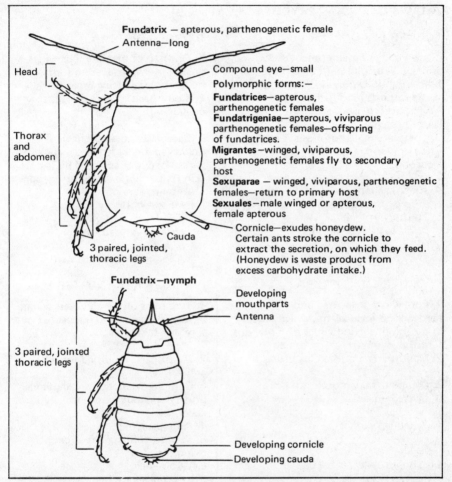

Fundatrix — apterous, parthenogenetic female
Antenna—long

Head

Compound eye—small
Polymorphic forms:—
Fundatrices—apterous, parthenogenetic females
Fundatrigeniae—apterous, viviparous parthenogenetic females—offspring of fundatrices.
Migrantes —winged, viviparous, parthenogenetic females fly to secondary host
Sexuparae — winged, viviparous, parthenogenetic females—return to primary host
Sexuales—male winged or apterous, female apterous

Thorax and abdomen

Cornicle—exudes honeydew. Certain ants stroke the cornicle to extract the secretion, on which they feed. (Honeydew is waste product from excess carbohydrate intake.)

Cauda

3 paired, jointed, thoracic legs

Fundatrix—nymph

Developing mouthparts
Antenna

3 paired, jointed thoracic legs

Developing cornicle
Developing cauda

Greenflies are found on young shoots and the foliage of plants on which they feed by piercing the plant tissue and sucking up the plant juices. Certain ant species tend herds of greenfly for the honeydew which they exude and on which the ants feed.

The winter is passed as eggs, laid in the previous autumn on a woody plant (primary host) by sexual females. The eggs hatch in the spring to form nymphs which develop into wingless parthenogenetic females, fundatrices, which produce several generations of similar wingless females, fundatrigeniae and some winged females, the migrantes which are concerned with migration and dispersal. If they find a suitable herbaceous plant (secondary host), they continue to reproduce parthenogenetically until the autumn when they produce a generation of winged sexuparae which return to the original primary host and produce the males and females, the sexuales. After mating, eggs are laid and the cycle is complete.

B) DIVISION ENDOPTERYGOTA (holometabola)

I ORDER LEPIDOPTERA

1. Two pairs of membranous scaly wings, colourfully patterned. Fore and hind wings joined by hooks.
2. Mouthparts modified to form coiled sucking proboscis. Mandibles rarely present.
3. Complete metamorphosis. Larvae often polypodous, pupae usually in a cocoon.

4. Antennae long.
e.g. *Pieris*—Cabbage white butterfly
Bombyx—silkworm moth

II ORDER DIPTERA

1. One pair of membranous wings, posterior pair reduced to form halteres or balancing organs.
2. Mouthparts modified for piercing or sucking.

3. Complete metamorphosis. Larvae apodous, terrestrial aquatic or parasitic, pupae enclosed by last larva skin (puparium) usually no cocoon.
4. Antennae variable.
e.g. *Musca*—housefly
Tipula—crane fly
Drosophila—fruitfly
Culex—gnat
Anopheles—mosquito

III ORDER COLEOPTERA

1. Fore wings thickened and hardened to form elytra, which meet in the midline. Hind wings membranous, folded beneath the elytra or absent.
2. Mouthparts modified for biting.

3. Complete metamorphosis, larvae of diverse types.

4. Antennae variable.

e.g. *Dytiscus*—water beetle

IV ORDER HYMENOPTERA

1. Two pairs of membranous wings, hind pair smaller and connected to fore pair by hooks.

2. Mouthparts modified for biting and licking.
3. Complete metamorphosis, larvae usually apodous, pupae usually in cocoons.
4. Antennae long.
5. Sawing, stinging or piercing ovipositor.
e.g. *Apis*—bee
Formica—ant

V ORDER TRICHOPTERA

1. Two pairs of membranous wings densely covered with hair.
2. Mouthparts reduced, mandibles not functional.
3. Complete metamorphosis, larvae aquatic, usually in portable cases, pupae aquatic with strong mandibles.
4. Antennae long.

e.g. *Limnephilus*—caddis fly

VI ORDER SIPHONAPTERA

1. Apterous, wings lost.

2. Mouthparts modified for piercing and sucking.
3. Complete metamorphosis, larvae apodous, pupae in silken cocoons.

4. Antennae short three segmented.
5. Laterally compressed, minute adults, ectoparasites with legs terminating in two claws.
e.g. *Pulex*—flea

Pieris
Cabbage white butterfly

PHYLUM ARTHROPODA

CLASS INSECTA

ORDER LEPIDOPTERA

GENUS *Pieris*

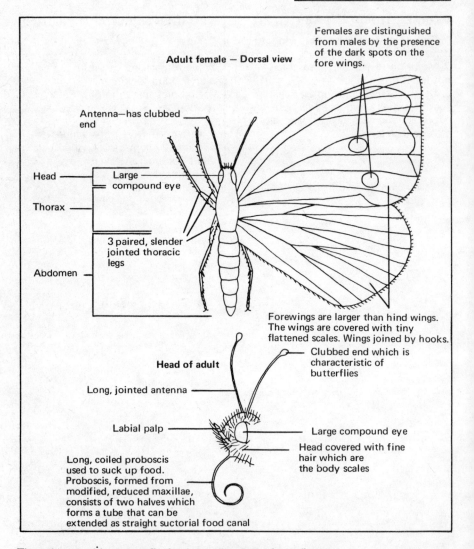

Adult female — Dorsal view

Females are distinguished from males by the presence of the dark spots on the fore wings.

Antenna—has clubbed end

Head

Large compound eye

Thorax

3 paired, slender jointed thoracic legs

Abdomen

Forewings are larger than hind wings. The wings are covered with tiny flattened scales. Wings joined by hooks.

Head of adult

Clubbed end which is characteristic of butterflies

Long, jointed antenna

Labial palp

Large compound eye

Head covered with fine hair which are the body scales

Long, coiled proboscis used to suck up food. Proboscis, formed from modified, reduced maxillae, consists of two halves which forms a tube that can be extended as straight suctorial food canal

The cabbage white butterfly feeds on the nectar from flowers.

Eggs are laid by the female on the under surface of cabbage leaves during the summer. The eggs hatch out into larvae or caterpillars which feed voraciously on the cabbage leaves and moult four times. When fully grown, the larva ceases to feed and wanders off in search of a suitable, sheltered place in which to pupate. On finding this, the pupa spins a small quantity of silk by which it is suspended. In the summer pupation may only take a few weeks but the animal may over winter in this state. The pupa metamorphoses into the adult form.

Pieris
Cabbage white butterfly

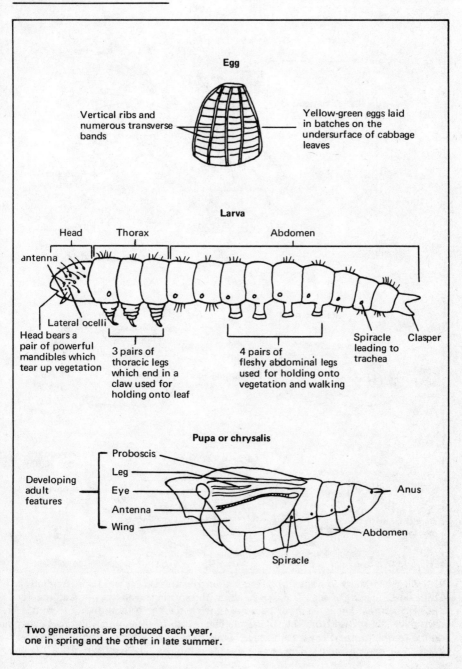

Egg

Vertical ribs and numerous transverse bands

Yellow-green eggs laid in batches on the undersurface of cabbage leaves

Larva

Head   Thorax   Abdomen

antenna

Lateral ocelli

Head bears a pair of powerful mandibles which tear up vegetation

3 pairs of thoracic legs which end in a claw used for holding onto leaf

4 pairs of fleshy abdominal legs used for holding onto vegetation and walking

Spiracle leading to trachea   Clasper

Pupa or chrysalis

Developing adult features

Proboscis
Leg
Eye
Antenna
Wing

Anus

Abdomen

Spiracle

Two generations are produced each year, one in spring and the other in late summer.

Bombyx
Silk worm moth

PHYLUM ARTHROPODA

CLASS INSECTA

ORDER LEPIDOPTERA

GENUS *Bombyx*

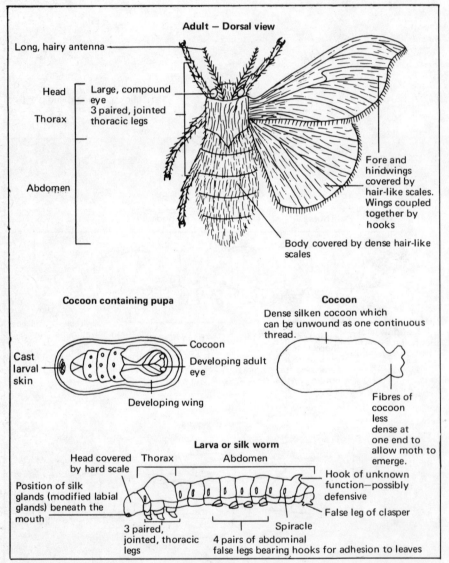

Adult – Dorsal view

Long, hairy antenna

Head

Large, compound eye

3 paired, jointed thoracic legs

Thorax

Abdomen

Fore and hindwings covered by hair-like scales. Wings coupled together by hooks

Body covered by dense hair-like scales

Cocoon containing pupa

Cast larval skin

Cocoon

Developing adult eye

Developing wing

Cocoon

Dense silken cocoon which can be unwound as one continuous thread.

Fibres of cocoon less dense at one end to allow moth to emerge.

Larva or silk worm

Head covered by hard scale

Thorax

Abdomen

Position of silk glands (modified labial glands) beneath the mouth

Hook of unknown function—possibly defensive

False leg of clasper

3 paired, jointed, thoracic legs

4 pairs of abdominal false legs bearing hooks for adhesion to leaves

Spiracle

The silk worm moth is native to oriental countries but is now bred commercially in China and Europe for silk. The eggs hatch in the spring and the larvae feed on mulberry leaves. The larva moults several times and when fully grown it spins a cocoon of silk thread from the spinneret. The length of time pupation takes depends on the temperature of the environment. Shortly after emergence from the cocoon as an adult mating occurs. The female then lays hundreds of eggs and dies.

Musca
Housefly

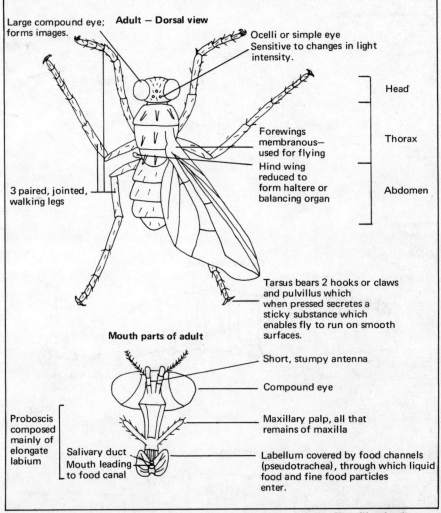

Adult — Dorsal view

Large compound eye; forms images.

Ocelli or simple eye Sensitive to changes in light intensity.

Head

Forewings membranous— used for flying

Thorax

Hind wing reduced to form haltere or balancing organ

Abdomen

3 paired, jointed, walking legs

Tarsus bears 2 hooks or claws and pulvillus which when pressed secretes a sticky substance which enables fly to run on smooth surfaces.

Mouth parts of adult

Short, stumpy antenna

Compound eye

Maxillary palp, all that remains of maxilla

Proboscis composed mainly of elongate labium

Salivary duct
Mouth leading to food canal

Labellum covered by food channels (pseudotrachea), through which liquid food and fine food particles enter.

House flies are found wherever dead organic matter is exposed. The flies feed on organic matter by first exuding saliva onto the food which dissolves it and then sucking up the resulting liquid.

During the summer months, eggs are laid in batches in decaying organic matter, The eggs hatch into white maggots which moult two or three times and pupate. The adult emerges from the pupal case within several days, but the pupa which are negatively phototactic may survive in the puparium over winter. Flies are of economic importance because they contaminate human food with pathogenic organisms.

57

Musca
Housefly

PHYLUM ARTHROPODA

CLASS INSECTA

ORDER DIPTERA

GENUS *Musca*

**Drawings on this page
are highly magnified**

Eggs

White, cigar-shaped
eggs which are laid in
batches of 100 eggs
during the summer
months.

Posterior
spiracles
lead to
trachea which
are used for
respiration

Larva or maggot

Anterior spiracles

Pads bearing setae
to facilitate movement

Hooked
mandible used
feeding and
locomotion

Puparium containing pupa

Anterior spiracle

Posterior
spiracle

Brown barrel shaped
structure

Tipula
Cranefly
or Daddy-longlegs

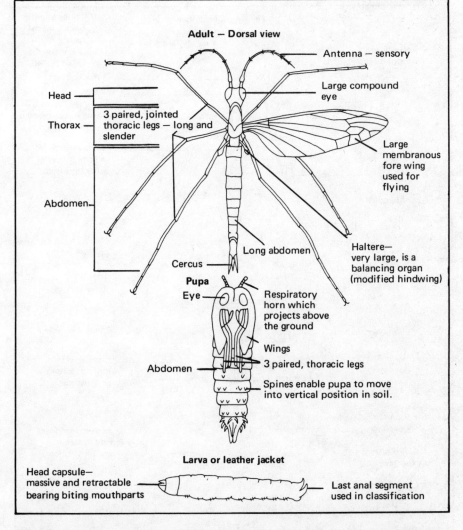

Adult — Dorsal view

Antenna — sensory

Head

Large compound eye

Thorax

3 paired, jointed thoracic legs — long and slender

Large membranous fore wing used for flying

Abdomen

Long abdomen

Haltere— very large, is a balancing organ (modified hindwing)

Cercus

Pupa

Eye

Respiratory horn which projects above the ground

Wings

Abdomen

3 paired, thoracic legs

Spines enable pupa to move into vertical position in soil.

Larva or leather jacket

Head capsule— massive and retractable bearing biting mouthparts

Last anal segment used in classification

Craneflies are found in environments varying from marshland to dry heathland. They feed rarely but are thought to feed on plant juices.

After mating, the female inserts her long abdomen into the ground and lays the long oval black eggs. The eggs hatch out into grey-brown larvae known as leather-jackets. These larvae are pests, they attack the roots and lower stems of cereals, grasses and other plants of economic importance. The larvae feed on this vegetation through the autumn, winter and spring, when they pupate. A few weeks later the adult cranefly emerges from the pupal case.

Drosophila
Fruit fly

PHYLUM ARTHROPODA

CLASS INSECTA

ORDER DIPTERA

GENUS *Drosophila*

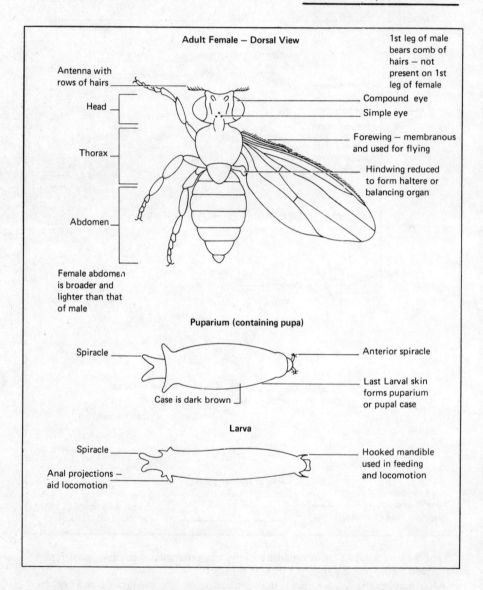

Adult Female — Dorsal View

1st leg of male bears comb of hairs — not present on 1st leg of female

Antenna with rows of hairs

Head

Thorax

Abdomen

Female abdomen is broader and lighter than that of male

Compound eye

Simple eye

Forewing — membranous and used for flying

Hindwing reduced to form haltere or balancing organ

Puparium (containing pupa)

Spiracle

Anterior spiracle

Last Larval skin forms puparium or pupal case

Case is dark brown

Larva

Spiracle

Anal projections — aid locomotion

Hooked mandible used in feeding and locomotion

The fruit fly is universal and is found wherever there are decaying fruits. After copulation, the female stores sperm and lays batches of white eggs which are deposited in the decaying fruit. The larvae are voracious feeders and active burrowers in the fruit. The pupae develop within the puparium and emerge as adults. The adult emerges in early morning, feeds and lives for about one month.

PHYLUM ARTHROPODA

CLASS INSECTA

ORDER DIPTERA

GENUS *Culex*

Culex
Gnat

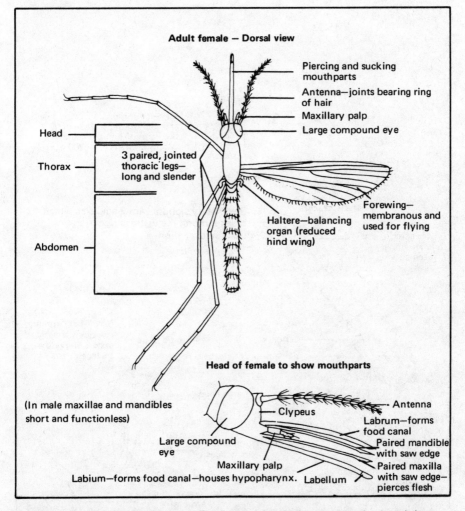

Adult female — Dorsal view

Piercing and sucking mouthparts

Antenna—joints bearing ring of hair

Maxillary palp

Large compound eye

Head

Thorax

3 paired, jointed thoracic legs— long and slender

Abdomen

Haltere—balancing organ (reduced hind wing)

Forewing— membranous and used for flying

Head of female to show mouthparts

(In male maxillae and mandibles short and functionless)

Antenna

Clypeus

Labrum—forms food canal

Large compound eye

Paired mandible with saw edge

Maxillary palp

Paired maxilla with saw edge— pierces flesh

Labium—forms food canal—houses hypopharynx. Labellum

Gnats are found in warm climates. The normal food of both sexes is plant juice which they obtain with the suctorial mouthparts, but the female has mouthparts which are modified for obtaining additional meals of blood.

The female lays eggs in rafts on the surface of stagnant water in ponds and ditches. Eggs hatch out into larvae which can swim actively. When at rest, the larvae hang up side down from the surface of the water. The larvae develop into free swimming pupae which do not feed. These too remain near the surface of the water when at rest. When metamorphosis is complete, the pupal case splits along the dorsal surface and the adult emerges onto the surface of the water. It rests on the pupal case until the wings are dry and flies away.

Culex
Gnat

PHYLUM ARTHROPODA

CLASS INSECTA

ORDER DIPTERA

GENUS *Culex*

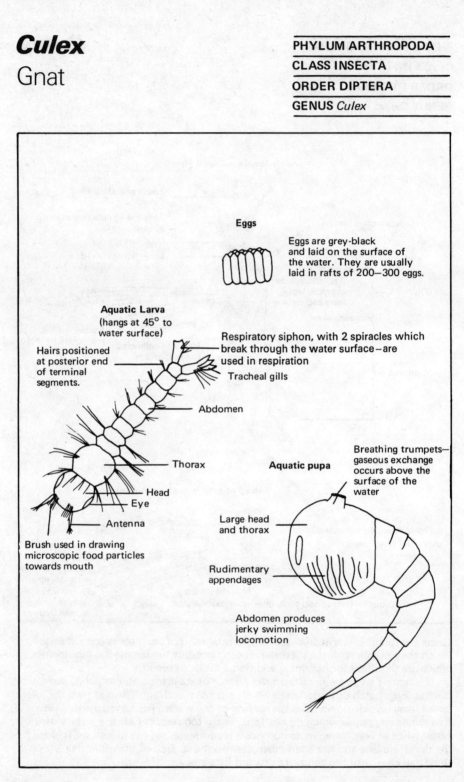

Eggs

Eggs are grey-black and laid on the surface of the water. They are usually laid in rafts of 200—300 eggs.

Aquatic Larva (hangs at 45° to water surface)

Hairs positioned at posterior end of terminal segments.

Respiratory siphon, with 2 spiracles which break through the water surface – are used in respiration

Tracheal gills

Abdomen

Thorax

Aquatic pupa

Head

Eye

Antenna

Brush used in drawing microscopic food particles towards mouth

Breathing trumpets— gaseous exchange occurs above the surface of the water

Large head and thorax

Rudimentary appendages

Abdomen produces jerky swimming locomotion

Anopheles
Mosquito

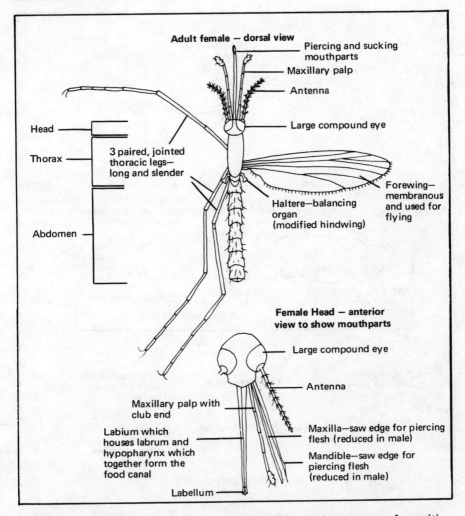

Adult female — dorsal view

Piercing and sucking mouthparts

Maxillary palp

Antenna

Large compound eye

Head

Thorax

3 paired, jointed thoracic legs— long and slender

Abdomen

Haltere—balancing organ (modified hindwing)

Forewing— membranous and used for flying

Female Head — anterior view to show mouthparts

Large compound eye

Antenna

Maxillary palp with club end

Labium which houses labrum and hypopharynx which together form the food canal

Maxilla—saw edge for piercing flesh (reduced in male)

Mandible—saw edge for piercing flesh (reduced in male)

Labellum

The mosquito is the carrier of the disease malaria, due to the presence of parasitic protozoans in the salivary gland, which are injected into man when the mosquito bites him.

The male feeds on plant juices, but the female is able to pierce the flesh of man into which she injects an anti-coagulant and sucks up blood.

After mating, the female lays eggs singly on the surface of the water. The eggs hatch out into larvae which are aquatic. The larvae moult several times and give rise to the pupae which are also aquatic. They swim freely, but do not feed. When the adult is fully developed within the pupa, the skin splits dorsally along the thoracic region and the adult emerges, rests in order to dry its wings and flies away.

Anopheles
Mosquito

PHYLUM ARTHROPODA

CLASS INSECTA

ORDER DIPTERA

GENUS *Anopheles*

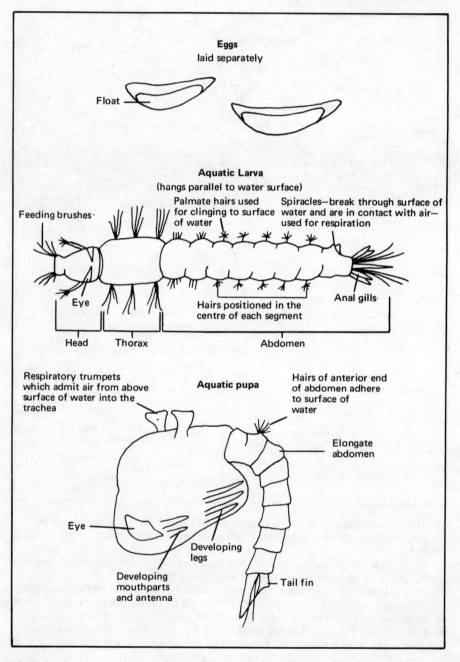

Eggs
laid separately

Float

Aquatic Larva
(hangs parallel to water surface)

Feeding brushes·

Palmate hairs used for clinging to surface of water

Spiracles—break through surface of water and are in contact with air—used for respiration

Eye

Hairs positioned in the centre of each segment

Anal gills·

Head   Thorax   Abdomen

Respiratory trumpets which admit air from above surface of water into the trachea

Aquatic pupa

Hairs of anterior end of abdomen adhere to surface of water

Elongate abdomen

Eye

Developing legs

Developing mouthparts and antenna

Tail fin

Dytiscus
Water beetle

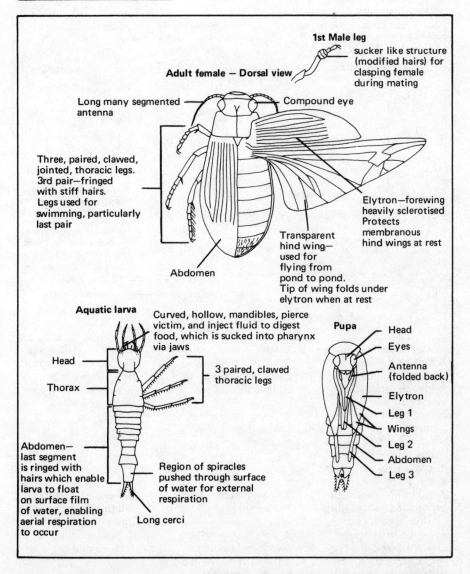

1st Male leg
sucker like structure
(modified hairs) for
clasping female
during mating

Adult female — Dorsal view

Long many segmented
antenna

Compound eye

Three, paired, clawed,
jointed, thoracic legs.
3rd pair—fringed
with stiff hairs.
Legs used for
swimming, particularly
last pair

Elytron—forewing
heavily sclerotised
Protects
membranous
hind wings at rest

Transparent
hind wing—
used for
flying from
pond to pond.
Tip of wing folds under
elytron when at rest

Abdomen

Aquatic larva

Curved, hollow, mandibles, pierce
victim, and inject fluid to digest
food, which is sucked into pharynx
via jaws

Pupa

Head

Eyes

Head

Antenna
(folded back)

Thorax

3 paired, clawed
thoracic legs

Elytron

Leg 1

Wings

Leg 2

Abdomen

Leg 3

Abdomen—
last segment
is ringed with
hairs which enable
larva to float
on surface film
of water, enabling
aerial respiration
to occur

Region of spiracles
pushed through surface
of water for external
respiration

Long cerci

Water beetles are freshwater animals. The adult and larva are carnivorous and both
respire through spiracles concentrated near tail end. The adult is also able to trap air
underneath the elytra. The female lays eggs inside the stems of water plants. The
egg hatches into a larva which preys on soft bodied animals. The water beetle
pupates in the moist earth near to the pond, as a pupa it does not feed. When the
adult beetle is formed the pupal case splits and the adult emerges.

Apis
Honey bee

PHYLUM ARTHROPODA

CLASS INSECTA

ORDER HYMENOPTERA

GENUS *Apis*

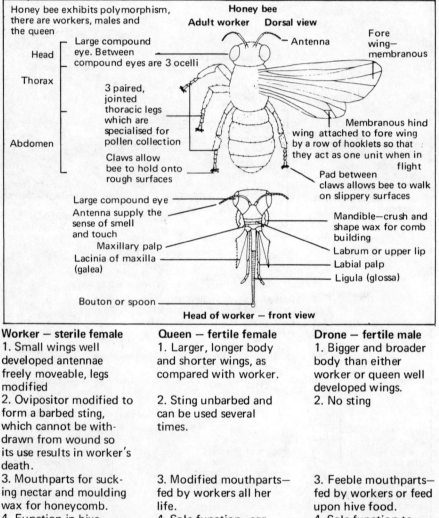

Honey bee exhibits polymorphism, there are workers, males and the queen

Honey bee
Adult worker   Dorsal view

Head

Thorax

Abdomen

Large compound eye. Between compound eyes are 3 ocelli

3 paired, jointed thoracic legs which are specialised for pollen collection

Claws allow bee to hold onto rough surfaces

Antenna

Fore wing— membranous

Membranous hind wing attached to fore wing by a row of hooklets so that they act as one unit when in flight

Pad between claws allows bee to walk on slippery surfaces

Large compound eye
Antenna supply the sense of smell and touch
Maxillary palp
Lacinia of maxilla (galea)

Bouton or spoon

Mandible—crush and shape wax for comb building
Labrum or upper lip
Labial palp
Ligula (glossa)

Head of worker — front view

Worker — sterile female
1. Small wings well developed antennae freely moveable, legs modified
2. Ovipositor modified to form a barbed sting, which cannot be withdrawn from wound so its use results in worker's death.
3. Mouthparts for sucking nectar and moulding wax for honeycomb.
4. Function in hive—produce wax cells, feed larvae, store food, clean and ventilate hive, and later collect pollen and nectar.
5. Short arduous life of 4—5 weeks.

Queen — fertile female
1. Larger, longer body and shorter wings, as compared with worker.

2. Sting unbarbed and can be used several times.

3. Modified mouthparts— fed by workers all her life.
4. Sole function—egg laying and leading a swarm from hive in order to set up new colony.

5. May live 5—6 years.

Drone — fertile male
1. Bigger and broader body than either worker or queen well developed wings.
2. No sting

3. Feeble mouthparts— fed by workers or feed upon hive food.
4. Sole function to fertilise queen.

5. Killed by workers who drive them out of the hive in the autumn.

PHYLUM ARTHROPODA
CLASS INSECTA
ORDER HYMENOPTERA
GENUS *Apis*

Apis
Honey bee

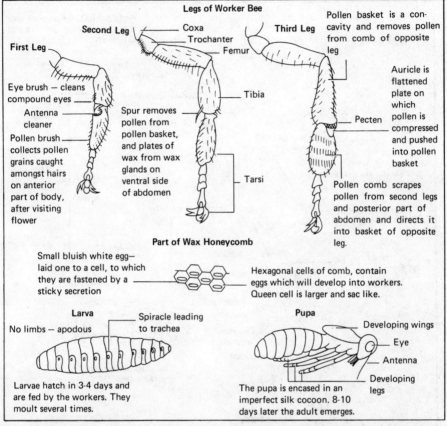

Legs of Worker Bee

First Leg

Eye brush — cleans compound eyes

Antenna cleaner

Pollen brush collects pollen grains caught amongst hairs on anterior part of body, after visiting flower

Second Leg

Coxa
Trochanter
Femur

Spur removes pollen from pollen basket, and plates of wax from wax glands on ventral side of abdomen

Tibia

Tarsi

Third Leg

Pollen basket is a concavity and removes pollen from comb of opposite leg

Auricle is flattened plate on which pollen is compressed and pushed into pollen basket

Pecten

Pollen comb scrapes pollen from second legs and posterior part of abdomen and directs it into basket of opposite leg.

Part of Wax Honeycomb

Small bluish white egg— laid one to a cell, to which they are fastened by a sticky secretion

Hexagonal cells of comb, contain eggs which will develop into workers. Queen cell is larger and sac like.

Larva

No limbs — apodous

Spiracle leading to trachea

Larvae hatch in 3-4 days and are fed by the workers. They moult several times.

Pupa

Developing wings
Eye
Antenna
Developing legs

The pupa is encased in an imperfect silk cocoon. 8-10 days later the adult emerges.

The honey bee is a social insect. It lives in a hive in a community which consists of a queen, several hundred drones and about 70,000 workers. The nuptial flight occurs in May when the young Queen leaves the hive and is fertilised in flight by only one drone. She returns to the hive and spends all her time laying eggs in the hexagonal wax cells which are made by the workers. Fertilised eggs produce female larvae. Those which are fed solely on royal jelly produce potential queens. Those which after a few days are fed on honey and pollen develop into workers. Unfertilised eggs give rise to drones.

The queen secretes a substance which the workers lick from her. This secretion prevents the workers from constructing a queen cell in the honeycomb. If the queen dies and with her the source of queen substance, the inhibition is lost and young queens appear in the colony. They duel until one emerges the victor. If overcrowding occurs in the colony, the queen substance becomes so diluted in being shared by such a large population that its effect is no longer felt everywhere in the hive. Some workers start queen raising activities, the colony splits in two, half go with the old queen and half remain in the old hive.

Formica
Ant

PHYLUM ARTHROPODA

CLASS INSECTA

ORDER HYMENOPTERA

GENUS *Formica*

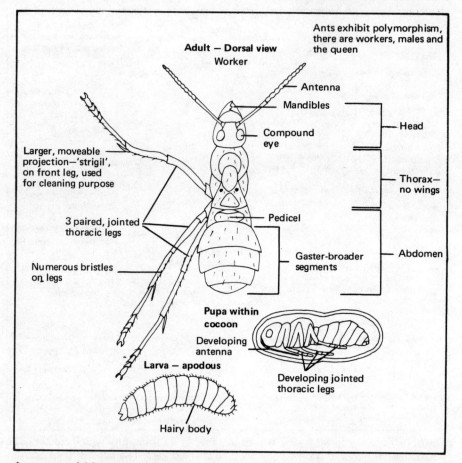

Ants exhibit polymorphism, there are workers, males and the queen

Adult — Dorsal view
Worker

Antenna

Mandibles

Compound eye

Head

Larger, moveable projection—'strigil', on front leg, used for cleaning purpose

Thorax— no wings

3 paired, jointed thoracic legs

Pedicel

Numerous bristles on legs

Gaster-broader segments

Abdomen

Pupa within cocoon

Developing antenna

Larva — apodous

Developing jointed thoracic legs

Hairy body

Ants are social insects which live in colonies in woody places. The colony consists of a queen, numerous female workers, and males. The ants feed on caterpillars, they also tend aphids for their honedew secretion, which they obtain by stroking the aphids' cornicles with their antennae.

The nuptial flight occurs in June, after which the fertilised queen returns to the nest, loses her wings and commences her role of egg laying. Meanwhile the winged males die. The queen lays eggs anywhere in the nest and they are picked up by the workers and carried off to the nurseries. Fertilised eggs give rise to workers and queens, and unfertilised eggs give rise to males. The eggs hatch out as legless larvae which are fed on regurgitated food from the workers. When the larvae are fully grown they spin a cocoon in which they pupate. When metamorphosis is complete, the adult emerges from the cocoon.

A new colony may be set up by the queen after her nuptial flight, but she must do all the work herself until she has reared a new generation of workers.

Limnephilus
Caddis fly

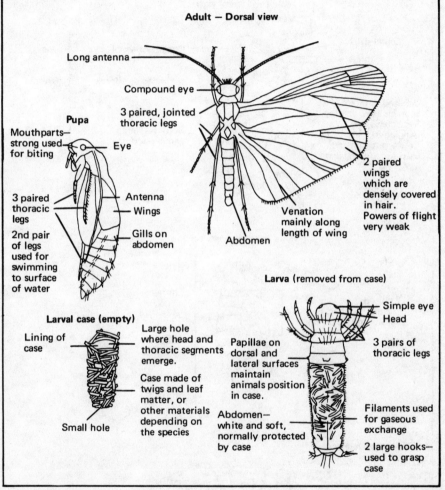

Adult — Dorsal view

Long antenna

Compound eye

3 paired, jointed thoracic legs

Pupa

Mouthparts—strong used for biting

Eye

3 paired thoracic legs

2nd pair of legs used for swimming to surface of water

Antenna

Wings

Gills on abdomen

2 paired wings which are densely covered in hair. Powers of flight very weak

Venation mainly along length of wing

Abdomen

Larva (removed from case)

Simple eye

Head

3 pairs of thoracic legs

Larval case (empty)

Lining of case

Large hole where head and thoracic segments emerge.

Case made of twigs and leaf matter, or other materials depending on the species

Small hole

Papillae on dorsal and lateral surfaces maintain animals position in case.

Abdomen—white and soft, normally protected by case

Filaments used for gaseous exchange

2 large hooks—used to grasp case

Caddis flies may be nocturnal or diurnal and are found in the vicinity of water. They have a dull colouring of browns and greys. Caddis flies have reduced mouthparts and are liquid feeders.

Eggs are laid in masses in or near water and develop into aquatic larvae which respire in the water by means of filamentous outgrowths of the abdomen. The larva seals the case by means of spun threads on pupating and the pupa remains quiescent. The pupa has strong jaws and is able to bite its way out of the case. It then swims to the surface of the water and climbs up a plant out of the water. The pupal skin splits and the adult emerges.

69

Pulex
Flea

PHYLUM ARTHROPODA

CLASS INSECTA

ORDER SIPHONAPTERA

GENUS *Pulex*

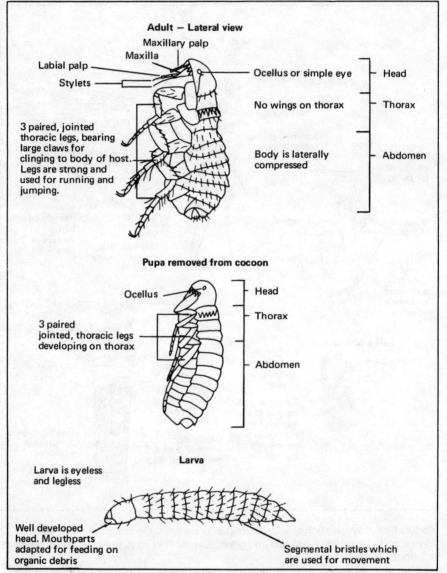

Adult — Lateral view

Maxillary palp

Maxilla

Labial palp

Stylets

Ocellus or simple eye — Head

No wings on thorax — Thorax

3 paired, jointed thoracic legs, bearing large claws for clinging to body of host. Legs are strong and used for running and jumping.

Body is laterally compressed — Abdomen

Pupa removed from cocoon

Ocellus

Head

Thorax

3 paired jointed, thoracic legs developing on thorax

Abdomen

Larva

Larva is eyeless and legless

Well developed head. Mouthparts adapted for feeding on organic debris

Segmental bristles which are used for movement

The common flea is ectoparasitic on man and has mouthparts adapted for piercing the skin and sucking blood. The rat flea transmits bacteria causing bubonic plague.

The female lays her eggs amongst the hair of the host, but they soon drop off and hatch out into larvae which feed for a couple of weeks. The larvae pupate inside silken cocoons and the adults emerge within another couple of weeks.

IV Class Arachnida

Usually terrestrial — no antennae.
Free living and parasitic.

External Features

Body divided into prosoma and opisthosoma.
Prosoma or cephalothorax consists of six adult
segments, bearing one pair of prehensile chelicerae,
one pair pedipalps and four pairs of walking legs.
Opisthosoma or abdomen of thirteen segments,
usually appendages are absent.
Simple eyes only.

Internal Features

Respiration by means of lung books and trachea.
Sexes usually separate.

Orders

SCORPIONIDEA

1. Prosoma covered by dorsal carapace.

2. Opisthosoma divided into mesosoma and metasoma bearing terminal sting (telson)

3. Chelicerae and pedipalps both prehensile.

4. Spinnerets absent.

5. Viviparous—young born alive.

e.g. *Scorpio*—
    Scorpion

ARANEIDA

1. Prosoma covered by single tergal shield, head marked off by a groove.

2. Opisthosoma soft, separated by a waist, from prosoma—no external segmentation.

3. Chelicerae prehensile, inject paralysing poison into prey. Pedipalps sensory modified for sperm transfer in the male.

4. Three pairs of spinnerets are present on the tip of the opisthosoma for spinning the web.

5. Direct development young hatch as miniature adults.

e.g. *Epeira*—
    garden spider

ACARINA

1. No boundary between prosoma and opisthosoma—bodies rounded.

2. —

3. Chelicerae prehensile, pedipalps united behind the mouth.

4. Spinnerets absent.

5. Eggs may hatch as "larvae" with three pairs of legs, or maybe viviparous.
e.g. *Ixodes*—tick
    Tyrophagus—mite

References for Class Arachnida

Savory, T.   1964   Spiders and Other Arachnids   London

Suggested Films for Class Arachnida

Gaumont British   1939   Arachnida (20 mins)   Black and White
Instructional

National Committee   1953   The Common Cross Spider   Black and White
For Visual Aids in   (10 mins)
Education

Scorpio
Scorpion

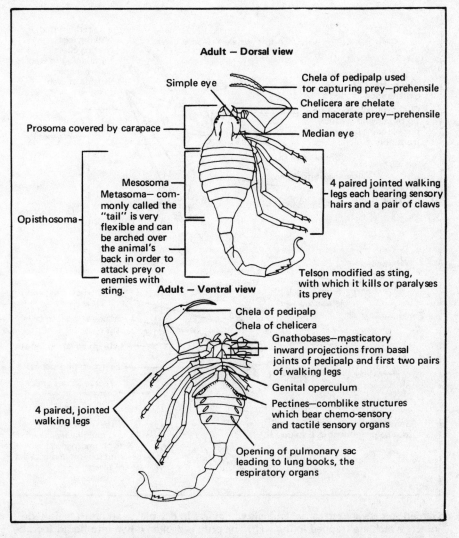

Adult — Dorsal view

Simple eye

Chela of pedipalp used for capturing prey—prehensile

Chelicera are chelate and macerate prey—prehensile

Prosoma covered by carapace

Median eye

Mesosoma

Metasoma— commonly called the "tail" is very flexible and can be arched over the animal's back in order to attack prey or enemies with sting.

Opisthosoma

4 paired jointed walking legs each bearing sensory hairs and a pair of claws

Telson modified as sting, with which it kills or paralyses its prey

Adult — Ventral view

Chela of pedipalp

Chela of chelicera

Gnathobases—masticatory inward projections from basal joints of pedipalp and first two pairs of walking legs

Genital operculum

Pectines—comblike structures which bear chemo-sensory and tactile sensory organs

4 paired, jointed walking legs

Opening of pulmonary sac leading to lung books, the respiratory organs

Scorpions are terrestrial animals and are confined to tropical and subtropical areas. They hide in crevices or beneath stones during the day and emerge at night to feed on invertebrates, particularly insects and spiders.

Courtship is elaborate in order to ensure that the female receives the spermatophore which the male has deposited on the ground. The male manoevres the female carefully so that the spermatophore, bearing a drop of sperm, enters her genital opening. The young are usually born alive (viviparity). The mother exhibits parental care by carrying the young on her back for several days.

Epeira
Garden spider

PHYLUM ARTHROPODA

CLASS ARACHNIDA

ORDER ARANEIDA

GENUS *Epeira*

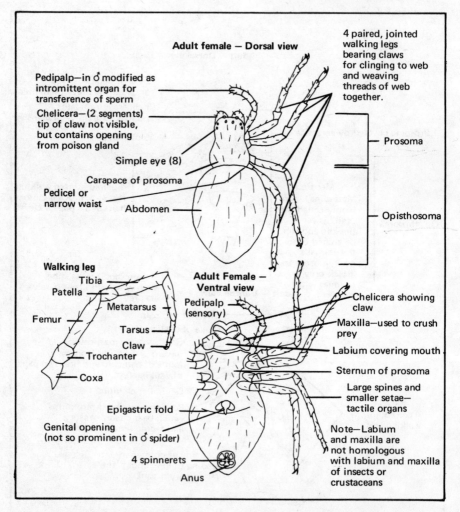

Adult female — Dorsal view

4 paired, jointed walking legs bearing claws for clinging to web and weaving threads of web together.

Pedipalp—in ♂ modified as intromittent organ for transference of sperm

Chelicera—(2 segments) tip of claw not visible, but contains opening from poison gland

Simple eye (8)

Carapace of prosoma

Pedicel or narrow waist

Abdomen

Prosoma

Opisthosoma

Walking leg

Tibia

Patella

Metatarsus

Femur

Tarsus

Claw

Trochanter

Coxa

Adult Female — Ventral view

Pedipalp (sensory)

Epigastric fold

Genital opening (not so prominent in ♂ spider)

4 spinnerets

Anus

Chelicera showing claw

Maxilla—used to crush prey

Labium covering mouth

Sternum of prosoma

Large spines and smaller setae— tactile organs

Note—Labium and maxilla are not homologous with labium and maxilla of insects or crustaceans

Garden spiders are terrestrial animals which live in dry places and feed mainly on insects which are trapped in the web. The prey is broken down into liquid food by secretions and sucked up by means of the muscular stomach. Male and female spiders are indistinguishable when immature and both spin webs. When sexually mature, the male spins a small sperm web, deposits on to it a drop of semen and dips the pedipalps into it to charge them with sperm.

An elaborate courtship precedes copulation during which the male inserts the pedipalps into the female genital opening. After copulation the female may kill and eat the male. Eggs are laid in silk cocoons and young spiders emerge as miniature adults. They undergo a series of moults before they reach sexual maturity.

PHYLUM ARTHROPODA
CLASS ARACHNIDA
ORDER ACARINA
GENUS *Ixodes*

Ixodes
Sheep tick

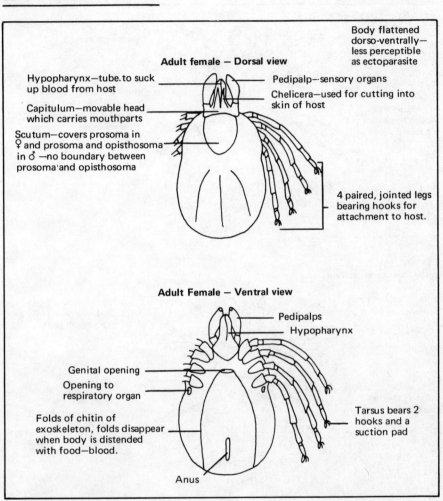

Body flattened dorso-ventrally— less perceptible as ectoparasite

Adult female — Dorsal view

Hypopharynx—tube.to suck up blood from host

Capitulum—movable head which carries mouthparts

Scutum—covers prosoma in ♀ and prosoma and opisthosoma in ♂ —no boundary between prosoma and opisthosoma

Pedipalp—sensory organs

Chelicera—used for cutting into skin of host

4 paired, jointed legs bearing hooks for attachment to host.

Adult Female — Ventral view

Pedipalps

Hypopharynx

Genital opening

Opening to respiratory organ

Folds of chitin of exoskeleton, folds disappear when body is distended with food—blood.

Tarsus bears 2 hooks and a suction pad

Anus

Ticks are ectoparasitic, blood sucking animals found on reptiles, birds and mammals. They may also transmit disease.

The female sheep tick takes a meal of blood and when fully gorged drops to the ground and lays eggs which hatch out into 6 legged "larvae". The "larvae" crawl up plant stems and are brushed off by a passing mammal on whose blood they feed. After feeding they drop off this first host and moult and emerge as a 8-legged nymph. These also climb vegetation and are brushed off by a second host on which they feed. They then drop off and moult into the adult form and climb up vegetation for the last time to find the third host. Adult male ticks do not feed until they find an adult female tick with whom they copulate. After copulation, the female feeds, drops to the ground, lays her eggs and then dies.

Tyrophagus
Mite

PHYLUM ARTHROPODA

CLASS ARACHNIDA

ORDER ACARINA

GENUS *Tyrophagus*

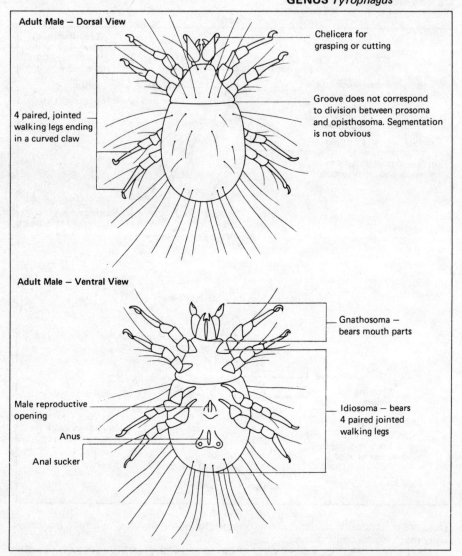

Adult Male — Dorsal View

Chelicera for grasping or cutting

Groove does not correspond to division between prosoma and opisthosoma. Segmentation is not obvious

4 paired, jointed walking legs ending in a curved claw

Adult Male — Ventral View

Gnathosoma — bears mouth parts

Idiosoma — bears 4 paired jointed walking legs

Male reproductive opening

Anus

Anal sucker

Mites of the genus *Tyrophagus* are world wide in distribution and feed largely upon stored food products with high protein and fat content. They feed mainly on fungi growing on these foods.

A typical life cycle begins with eggs which hatch out as 6-legged "larvae". The "larvae" moult into 8-legged nymphs which may moult several times before giving rise to the adult. There are many modifications of this life cycle. Certain nymphs moult to form a hypopus stage before developing into the adult. The hypopus is very small and is responsible for dispersal.

Phylum Mollusca

Ubiquitous distribution.
Free living and parasitic forms.
Triploblastic coelomates—show no segmentation.
Basic bilateral symmetry, but body may undergo
torsion resulting in asymmetry.

External Features

Soft body basically divided into head, foot and
visceral mass.
Calcareous shell may be present.
Mantle secretes shell.

Internal Features

Mantle encloses mantle cavity or pallial cavity, which may
house respiratory gills or ctenidia, openings of
gonoducts, kidneys and rectum.

CLASSES—

I GASTROPODA	II LAMELLIBRANCHIA (BIVALVIA)	III CEPHALOPODA
1. Exhibit asymmetry.	1. Exhibit bilateral symmetry.	1. Exhibit bilateral symmetry.
2. Slow moving molluscs.	2. Sedentary molluscs which may exhibit restricted movement.	2. Fast moving molluscs
3. Head well developed bears tentacles and eyes.	3. Head rudimentary, tentacles absent.	3. Head well developed and surrounded by well developed prehensile tentacles—modified foot.
4. Visceral mass undergoes torsion.	4. Visceral mass does not undergo torsion.	4. Visceral mass does not undergo torsion.
5. Flat foot well developed and used for "creeping".	5. Foot of variable size often used for burrowing.	5. Foot modified to form tentacles and siphon.
6. Radula or scraping tongue present in head for herbivorous and other feeding.	6. Two well developed ctenidia bear cilia used for filter feeding	6. Predacious carnivores using prehensile tentacles and horny jaws.
7. Mantle secretes shell in one piece, usually coiled.	7. Mantle in two lobes, secretes shell in two valves.	7. Large mantle may enclose reduced shell.
		8. Well developed pair of eyes.

e.g. *Helix*—snail
    Limax—slug
	e.g. *Mytilus*—mussel
	    Pecten—scallop
		e.g. *Sepia*—cuttlefish
		    Loligo—squid
		    Eledone—octopus

References for Phylum Mollusca

Lane, F.W.	1960	Kingdom of the Octopus	New York
Morton, J.E.	1964	Molluscs	London
Thorson, G.	1946	Reproduction and Larval Development of Danish Marine Bottom Invertebrates	Denmark

Suggested Films for Phylum Mollusca

Enclyclopaedia Britannica	1966	Life Story of the Snail (9 mins)	Colour

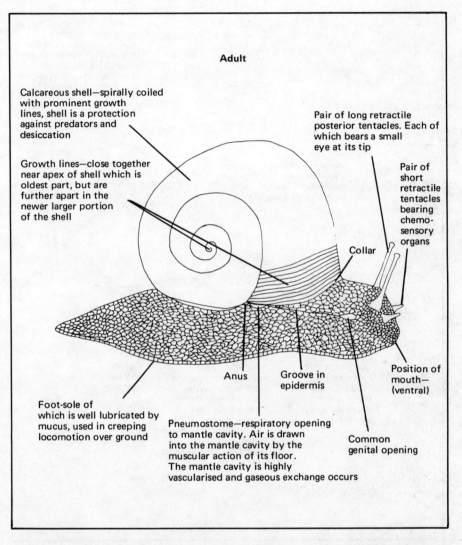

Adult

Calcareous shell—spirally coiled with prominent growth lines, shell is a protection against predators and desiccation

Growth lines—close together near apex of shell which is oldest part, but are further apart in the newer larger portion of the shell

Pair of long retractile posterior tentacles. Each of which bears a small eye at its tip

Pair of short retractile tentacles bearing chemo-sensory organs

Collar

Anus

Groove in epidermis

Position of mouth—(ventral)

Foot-sole of which is well lubricated by mucus, used in creeping locomotion over ground

Pneumostome—respiratory opening to mantle cavity. Air is drawn into the mantle cavity by the muscular action of its floor. The mantle cavity is highly vascularised and gaseous exchange occurs

Common genital opening

The snail is a terrestrial animal which feeds on vegetable matter, which it fragments by a toothed radula. It hibernates in winter by withdrawing the head and foot into the shell and sealing the shell aperture with a mucous secretion hardened with calcium salts.

Although the snail is hermaphrodite, there is a complicated copulatory process which gives rise to fertilised eggs. Although the eggs have a shell, which is mainly calcium phosphate, they are permeable to water so they are laid in damp situations. Development is direct and miniature adults emerge from the egg.

Limax
Slug

PHYLUM MOLLUSCA

CLASS GASTROPODA

GENUS *Limax*

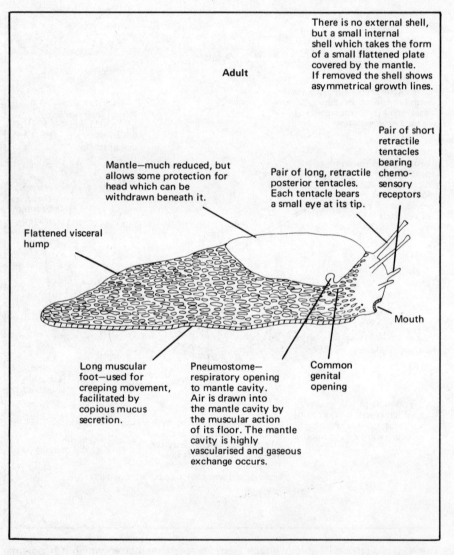

Adult

There is no external shell, but a small internal shell which takes the form of a small flattened plate covered by the mantle. If removed the shell shows asymmetrical growth lines.

Pair of short retractile tentacles bearing chemo-sensory receptors

Mantle—much reduced, but allows some protection for head which can be withdrawn beneath it.

Pair of long, retractile posterior tentacles. Each tentacle bears a small eye at its tip.

Flattened visceral hump

Mouth

Long muscular foot—used for creeping movement, facilitated by copious mucus secretion.

Pneumostome—respiratory opening to mantle cavity. Air is drawn into the mantle cavity by the muscular action of its floor. The mantle cavity is highly vascularised and gaseous exchange occurs.

Common genital opening

Slugs are commonly found on cultivated land and gardens. They are herbivores which emerge to feed in the evening and after rain, otherwise they live in dark, damp places since they have little protection against predators and desiccation. Sexual reproduction occurs when the slugs are at least two years old. Although they are hermaphrodite, cross fertilisation is usual, since they are protandrous. A complicated copulation gives rise to shelled eggs, which are permeable to water and so laid in damp situations. A miniature adult emerges from each egg—development is direct.

Mytilus
Mussel

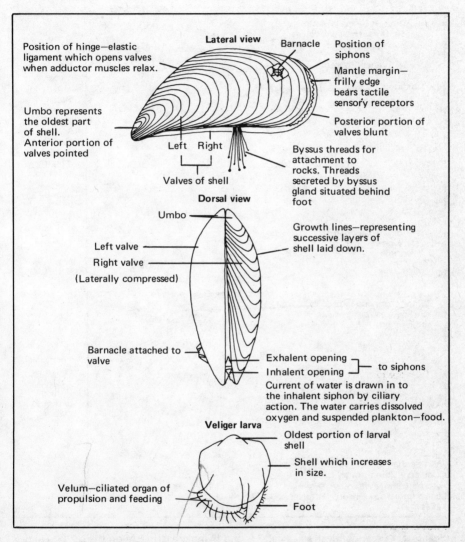

Lateral view

Position of hinge—elastic ligament which opens valves when adductor muscles relax.

Barnacle

Position of siphons

Mantle margin— frilly edge bears tactile sensory receptors

Umbo represents the oldest part of shell. Anterior portion of valves pointed

Posterior portion of valves blunt

Left   Right

Byssus threads for attachment to rocks. Threads secreted by byssus gland situated behind foot

Valves of shell

Dorsal view

Umbo

Left valve

Right valve

(Laterally compressed)

Growth lines—representing successive layers of shell laid down.

Barnacle attached to valve

Exhalent opening ⎤
Inhalent opening ⎦ to siphons

Current of water is drawn in to the inhalent siphon by ciliary action. The water carries dissolved oxygen and suspended plankton—food.

Veliger larva

Oldest portion of larval shell

Shell which increases in size.

Velum—ciliated organ of propulsion and feeding

Foot

Mussels are marine bivalves which live attached to rocks or piers by means of the byssus threads. They are filter feeders, extracting plankton from the current of water which is drawn in by the cilia on the ctenidia or gills.

Sexes are separate, gametes are released into the sea and fertilisation is external. The zygote develops into a trochophore larva, which develops into a second larval type, a veliger, which is a dominant plankton form at certain times of the year. When the larva is washed ashore, it becomes attached to the rocks and metamorphoses into the adult.

Pecten
Scallop

PHYLUM MOLLUSCA

CLASS LAMELLIBRANCHIA (BIVALVIA)

GENUS *Pecten*

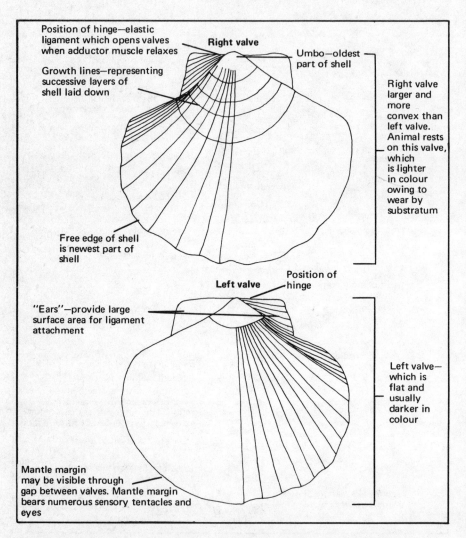

Position of hinge—elastic ligament which opens valves when adductor muscle relaxes

Right valve

Umbo—oldest part of shell

Growth lines—representing successive layers of shell laid down

Right valve larger and more convex than left valve. Animal rests on this valve, which is lighter in colour owing to wear by substratum

Free edge of shell is newest part of shell

Position of hinge

Left valve

"Ears"—provide large surface area for ligament attachment

Left valve—which is flat and usually darker in colour

Mantle margin may be visible through gap between valves. Mantle margin bears numerous sensory tentacles and eyes

Scallops are marine bivalves, usually found living gregariously on the sea bed and occasionally washed up on the lower shore. When very small they are attached by byssus threads to prevent being washed away by the current, later they swim by flapping the two valves. Most of their life is passed resting on the right valve while they feed. Water is drawn into the animal by cilia on the ctenidia or gills. The water current carries suspended plankton and dissolved oxygen.

Scallops are hermaphrodite. Gametes are liberated into the sea and fertilisation is external giving rise to a trochophore larva which develops into a second larva, the veliger, which metamorphoses into the adult.

Sepia
Cuttlefish

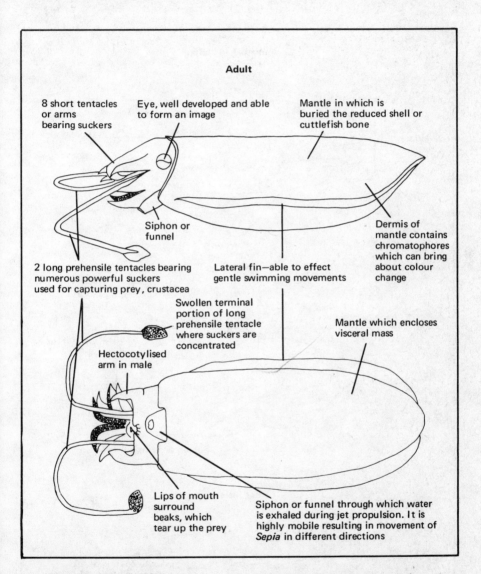

Adult

8 short tentacles or arms bearing suckers

Eye, well developed and able to form an image

Mantle in which is buried the reduced shell or cuttlefish bone

Siphon or funnel

2 long prehensile tentacles bearing numerous powerful suckers used for capturing prey, crustacea

Lateral fin—able to effect gentle swimming movements

Dermis of mantle contains chromatophores which can bring about colour change

Swollen terminal portion of long prehensile tentacle where suckers are concentrated

Hectocotylised arm in male

Mantle which encloses visceral mass

Lips of mouth surround beaks, which tear up the prey

Siphon or funnel through which water is exhaled during jet propulsion. It is highly mobile resulting in movement of *Sepia* in different directions

The cuttlefish is a marine animal, found close to the bottom of inshore waters. It is carnivorous in its diet, feeding on shrimps and prawns. It does not swim very quickly, but spends long periods resting on the sea floor, half covered by sand.

During copulation sperm enclosed in elastic tubes or spermatophores are transferred to the mantle cavity of the female by the hectocotylised arm which has suppressed suckers. The eggs are attached to seaweed by india-rubber like threads. The young hatch out as miniature adults.

Sepia
Cuttlefish

PHYLUM MOLLUSCA

CLASS CEPHALOPODA

GENUS *Sepia*

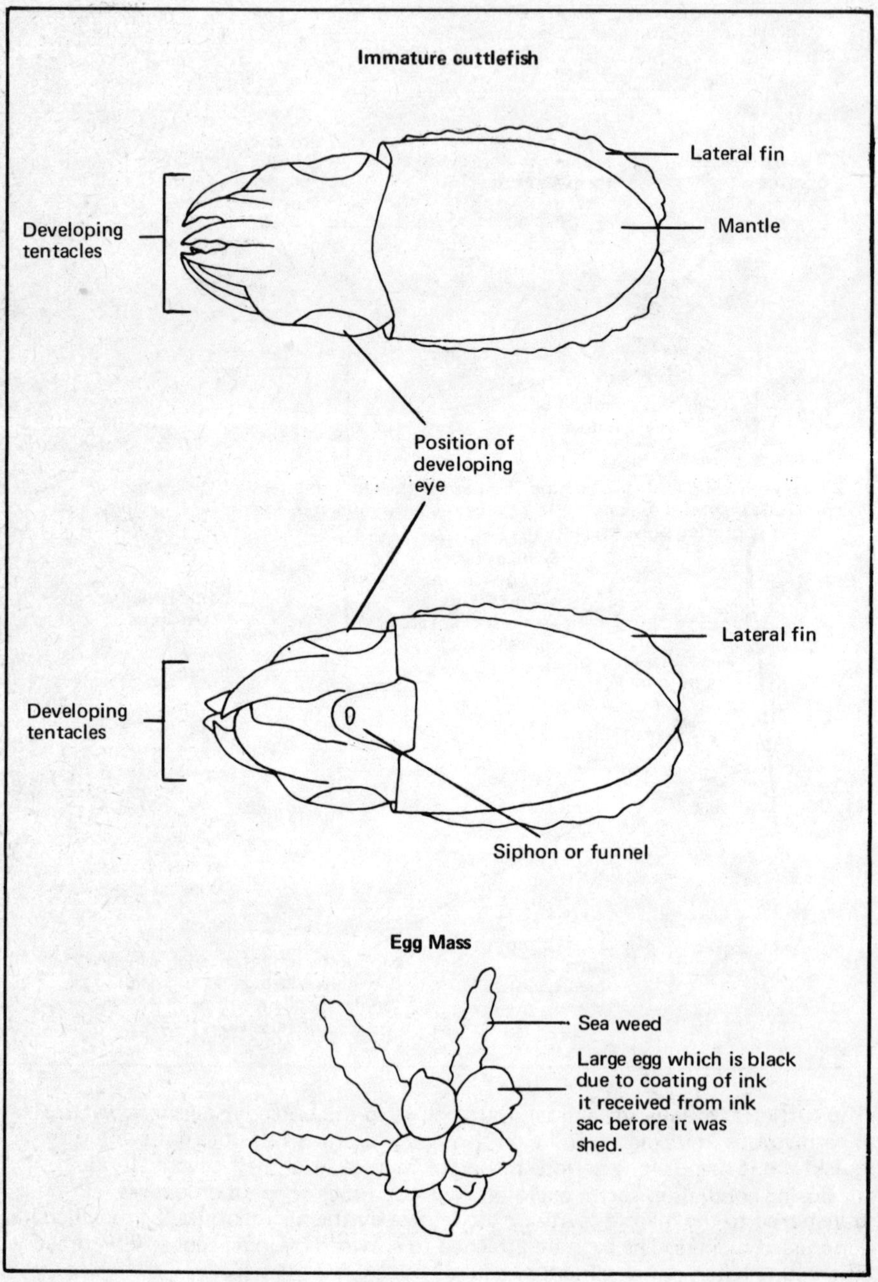

Immature cuttlefish

Lateral fin

Mantle

Developing tentacles

Position of developing eye

Lateral fin

Developing tentacles

Siphon or funnel

Egg Mass

Sea weed

Large egg which is black due to coating of ink it received from ink sac before it was shed.

PHYLUM MOLLUSCA

CLASS CEPHALOPODA

GENUS *Loligo*

Loligo
Squid

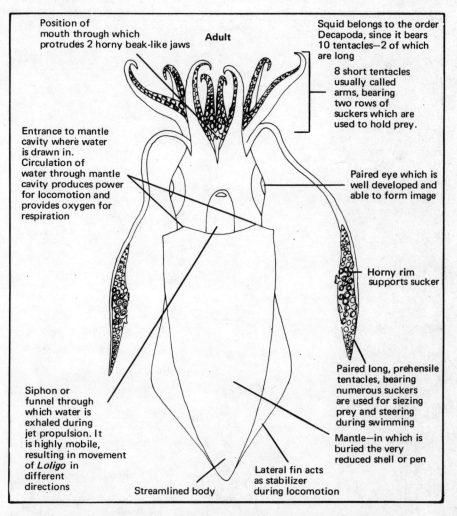

Position of mouth through which protrudes 2 horny beak-like jaws

Adult

Squid belongs to the order Decapoda, since it bears 10 tentacles—2 of which are long

8 short tentacles usually called arms, bearing two rows of suckers which are used to hold prey.

Entrance to mantle cavity where water is drawn in. Circulation of water through mantle cavity produces power for locomotion and provides oxygen for respiration

Paired eye which is well developed and able to form image

Horny rim supports sucker

Siphon or funnel through which water is exhaled during jet propulsion. It is highly mobile, resulting in movement of *Loligo* in different directions

Paired long, prehensile tentacles, bearing numerous suckers are used for siezing prey and steering during swimming

Mantle—in which is buried the very reduced shell or pen

Streamlined body

Lateral fin acts as stabilizer during locomotion

Squids are fast swimming marine animals. They are carnivorous, and prey on crustacea and fish which they seize using their suctorial tentacles while they proceed to bite with their horny jaws.

Fertilisation is internal. During copulation spermatophores are transferred to the female's mantle cavity by the hectocotylus of the male. The hectocotylus is the short arm on which several suckers are modified to form an attachment area for the spermatophores. Courtship and copulation are communal activities. Eggs, encased in buoyant pods of jelly, are ejected through the siphon. Other females add their egg pods to the communal nursery on the sea bottom. The young emerge from the eggs as miniature adults.

Loligo
Squid

PHYLUM MOLLUSCA

CLASS CEPHALOPODA

GENUS *Loligo*

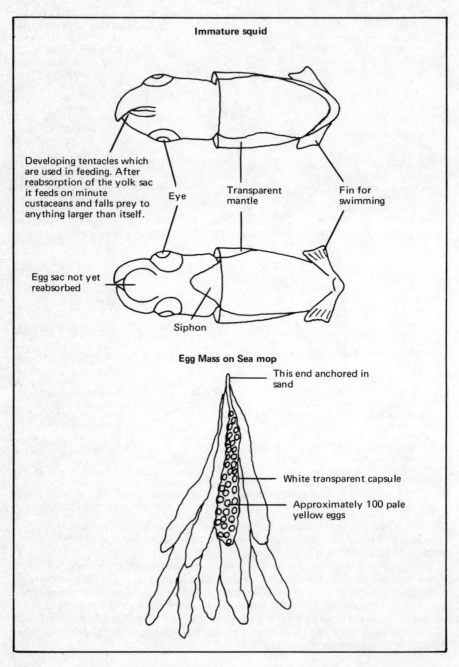

Immature squid

Developing tentacles which are used in feeding. After reabsorption of the yolk sac it feeds on minute custaceans and falls prey to anything larger than itself.

Eye

Transparent mantle

Fin for swimming

Egg sac not yet reabsorbed

Siphon

Egg Mass on Sea mop

This end anchored in sand

White transparent capsule

Approximately 100 pale yellow eggs

Eledone
Lesser octopus

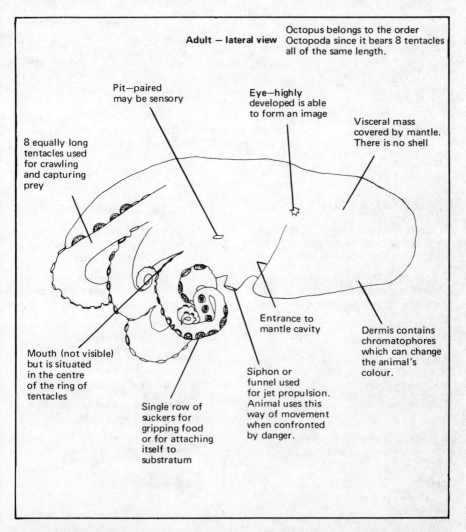

Adult — lateral view Octopus belongs to the order Octopoda since it bears 8 tentacles all of the same length.

Pit—paired may be sensory

Eye—highly developed is able to form an image

Visceral mass covered by mantle. There is no shell

8 equally long tentacles used for crawling and capturing prey

Mouth (not visible) but is situated in the centre of the ring of tentacles

Single row of suckers for gripping food or for attaching itself to substratum

Entrance to mantle cavity

Siphon or funnel used for jet propulsion. Animal uses this way of movement when confronted by danger.

Dermis contains chromatophores which can change the animal's colour.

The octopus is a predacious marine bottom dweller which hides under rocks and in crevices during the day and emerges at night to seek its prey, crabs and other crustacea. The octopus is able to swim, in a rather jerky way, but its usual means of movement is to crawl over the rocks using its long tentacles.

The male octopus has a hectocotylised tentacle which is spoon shaped at the end. This hectocotylus is used for spermatophore transfer during sexual reproduction when it is inserted into the mantle cavity of the female. The eggs are laid and resemble a bunch of grapes attached to the rocks. The octopus shows a certain amount of parental care until the eggs hatch out as miniature adults.

Phylum Echinodermata

Spiny skinned animals.
Marine and free living.
Triploblastic coelomates.
Adult pentamerously radially symmetrical.
Larva bilaterally symmetrical.

External Features

Spines, which may be movable, usually present.
Tube feet, which may end in suckers, present.

Internal Features

Nervous system consists of nerve ring around the mouth, giving off radial nerve cords.
Water vascular system derived from the coelom, concerned with the functions of the tube feet.
Sexes usually separate.

I CLASS ASTEROIDEA

1. Flattened, star shaped, ambulacral grooves open, border region of tube feet, restricted to oral surface.

2. Few calcareous plates or ossicles present, embedded in body wall and joined by connective tissue—flexible.

3. Various types of pincer-like structures, pedicellariae.

4. Mouth on oral surface, anus and opening to water vascular system, madreporite, on aboral surface.

5. No jaw apparatus.

6. Herbivorous, omnivorous or carnivorous.

7. Bipinnaria or brachiolaria larva.
e.g. *Asterias*—star fish

II CLASS ECHINOIDEA

1. Globular or discoid, ambulacral grooves closed—tube feet penetrate through pores in the plates.

2. Numerous calcareous plates present, attached rigidly to form corona or test.

3. Various types of pincer-like structures, pedicellariae with long stalks.

4. Mouth on oral surface, anus and opening to water vascular system, madreporite, on aboral surface.

5. Complicated jaw apparatus, showing pentamerous arrangement.

6. Herbivorous, omnivorous, sometimes carnivorous.

7. Echinopluteus larva.
e.g. *Echinus*—sea urchin

References for Phylum Echinodermata

Nichols, D.   1967   Echinoderms   London

Asterias
Starfish

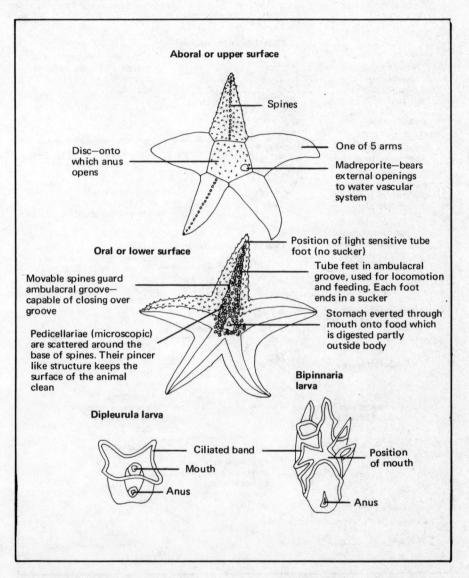

Aboral or upper surface

Spines

Disc—onto which anus opens

One of 5 arms

Madreporite—bears external openings to water vascular system

Position of light sensitive tube foot (no sucker)

Oral or lower surface

Tube feet in ambulacral groove, used for locomotion and feeding. Each foot ends in a sucker

Movable spines guard ambulacral groove—capable of closing over groove

Stomach everted through mouth onto food which is digested partly outside body

Pedicellariae (microscopic) are scattered around the base of spines. Their pincer like structure keeps the surface of the animal clean

Bipinnaria larva

Dipleurula larva

Ciliated band

Mouth

Anus

Position of mouth

Anus

Starfish are marine animals which live on the sea bottom at various depths. They are carnivorous, feeding mostly on bivalves which they open by applying the arms against the two valves, and using the tube feet, pull the valves open.

Sexes are separate and fertilisation occurs externally. The zygote gives rise to a dipleurula, which develops into a bipinnaria which results in a brachiolaria larva, which metamorphoses into an adult starfish.

Echinus
Sea urchin

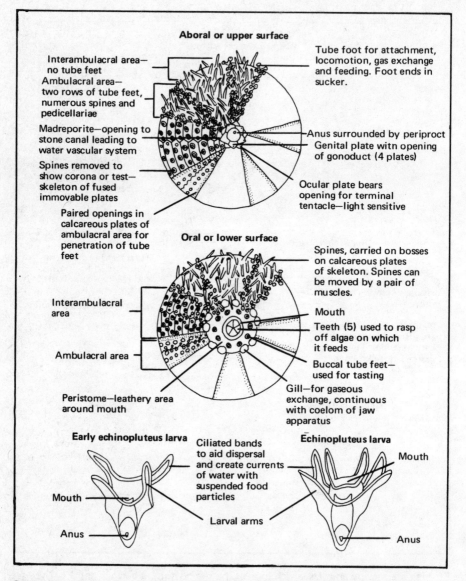

Aboral or upper surface

Interambulacral area—
no tube feet
Ambulacral area—
two rows of tube feet,
numerous spines and
pedicellariae

Madreporite—opening to
stone canal leading to
water vascular system

Spines removed to
show corona or test—
skeleton of fused
immovable plates

Paired openings in
calcareous plates of
ambulacral area for
penetration of tube
feet

Tube foot for attachment,
locomotion, gas exchange
and feeding. Foot ends in
sucker.

Anus surrounded by periproct
Genital plate with opening
of gonoduct (4 plates)

Ocular plate bears
opening for terminal
tentacle—light sensitive

Oral or lower surface

Interambulacral
area

Ambulacral area

Peristome—leathery area
around mouth

Spines, carried on bosses
on calcareous plates
of skeleton. Spines can
be moved by a pair of
muscles.

Mouth

Teeth (5) used to rasp
off algae on which
it feeds

Buccal tube feet—
used for tasting

Gill—for gaseous
exchange, continuous
with coelom of jaw
apparatus

Early echinopluteus larva

Mouth

Anus

Ciliated bands
to aid dispersal
and create currents
of water with
suspended food
particles

Larval arms

Echinopluteus larva

Mouth

Anus

Sea urchins are marine animals found on the sea bottom where there is a thick
growth of algae. They creep along by means of movable spines. Sexes are
separate and reproduction gives rise to a zygote which develops into a dipleurula
larva which develops into a second larval form, the echinopluteus larva which
undergoes metamorphosis to give the adult.

Experimental Approach to the Invertebrates

STAINING SCHEDULE FOR PERMANENT PREPARATIONS

A Borax Carmine
1 50% alcohol for 2 minutes
2 Stain in borax carmine for 5 to 10 minutes.
3 Wash in 50% alcohol.
4 Wash in 70% alcohol.
5 Differentiate in acid alcohol and check under the microscope.
6 Wash in 70% alcohol.
7 Rinse in 90% alcohol for 1 minute.
8 Transfer to absolute alcohol for two rinses of 5 minutes each. Keep covered.
9 Clear in xylol for at least 10 minutes.
10 Mount in Canada balsam.
NB Borax Carmine can be used to stain the coelenterate, *Obelia*. (The specimen should not be allowed to dry out in contact with the air because it will become darkened in appearance and brittle.)

B Haematoxylin
1 70% alcohol for 2 minutes.
2 Stain in haematoxylin for 5 to 15 minutes.
3 Rinse in 70% alcohol.
4 Differentiate in acid alcohol and check under the microscope.
5 Wash in 70% alcohol.
6 Run under tap water until blue (haematoxylin turns red in acid and blue in alkaline conditions).
7 Wash in 70% alcohol.
8 Counterstain in eosin for 10 seconds (this is optional).
9 Rinse in 90% alcohol for 1 minute.
10 Transfer to absolute alcohol for two rinses of 5 minutes each. Keep covered.
11 Clear in xylol for at least 10 minutes.
12 Mount in Canada balsam.
NB (i) Haematoxylin can be used to stain transverse sections of the annelid, *Lumbricus*. These preparations will be embedded in wax for sectioning purposes, and before staining the wax must be dissolved off with xylol. (Several rinses in xylol of 5 to 10 minutes each may be necessary to remove all the wax.) Then wash in 70% alcohol and proceed with the normal staining schedule.

(ii) This stain may also be used on a smear of the seminal vesicles of *Lumbricus*, which often contain stages in the life cycle of the protozoan parasite *Monocystis.*

PREPARATION OF A SMEAR OF THE SEMINAL VESICLES OF *LUMBRICUS*

1 Remove one seminal vesicle from a freshly dissected earthworm. (Dissect the animal dry and not under water.)
2 Squash it to a milky consistency at one end of a clean microscope slide, adding a drop of invertebrate saline to make the preparation more fluid, if necessary.
NB Preserved earthworms can also be used, but the tissues tend to be hardened and therefore difficult to reduce to a milky fluid.
3 Tilt the slide so that the milky fluid runs down the length of the slide, leaving the solid pieces at one end, and a thin smear over a large area of the slide.
4 Air dry the preparation for sufficient time to ensure that the smear firmly adheres to the slide and will not dissolve off.
5 Fix in absolute alcohol for about five minutes if a fresh earthworm is used and then follow the Haematoxylin staining schedule for the preparation of a permanent mount.

PERMANENT PREPARATION OF INSECT MOUTHPART WITHOUT STAINING

1 Cut off the head of a freshly killed specimen, e.g. locust, cockroach, bee, etc., and boil the whole head in 10% caustic soda (sodium hydroxide) until the tissues are softened.
NB When the head sinks to the bottom of the test tube it has been boiled sufficiently.
2 Remove with forceps and rinse thoroughly under tap water.
3 Sever all the mouthparts carefully from their point of attachment to the head and immerse in 70% alcohol immediately.
4 Dehydrate in 90% alcohol and then in absolute alcohol.
5 Clear in xylol.
6 Mount in Canada balsam and cover with a coverslip, having arranged the mouthparts in the centre of the slide.
NB Rather than lower the coverslip obliquely onto the preparation, causing the mouthparts to be displaced, put a drop of Canada balsam on the centre of the coverslip and place a few drops around the preparation on the microscope slide and then lower the coverslip vertically onto the slide.

INVESTIGATION OF FILTER FEEDING AND FOOD VACUOLE FORMATION IN *PARAMECIUM*

Apparatus
Paramecium culture (the specimens should be preferably large in size)
A solution of yeast stained with Congo red
A suspension of graphite particles
'Polycel' or glycerine
Cavity slides and coverslips
Pipettes
Microscope

Method
1 Place a drop of liquid containing *Paramecium* onto a cavity slide. Add a drop of glycerine and carefully lower a coverslip onto the preparation. Check under the microscope for the presence of the organism.
NB The addition of glycerine will help to reduce the activity of *Paramecium*, so that it can be observed more readily during the experiment.
2 Now place a drop of the culture and a drop of glycerine onto another cavity slide, add a drop of yeast stained in Congo red, cover with a coverslip and view the organisms under the microscope at low power.
NB Congo red is a pH indicator, which is red in colour in alkaline or neutral media, but blue in acid conditions.
3 Repeat, using a fresh sample of culture, but this time add a drop of a suspension of graphite particles.
4 Examine each of the three slides, using the high power objective of the microscope at regular intervals of 10, 20 and 30 minutes. Make careful observations, noting any differences in the second and third slides compared with the first control slide.

Questions
1 Did *Paramecium* take up the yeast stained in Congo red and was there any evidence for the yeast being enclosed in food vacuoles inside the organism?
2 Where did the food vacuoles occur in the animal over the 30-minute period?
3 Was there any observable colour change in the Congo red indicator during this time, and if so, why?
4 Similarly, was the suspension of graphite particles taken up by *Paramecium*, or was it rejected. If the former occurred, were these particles surrounded by food vacuoles or merely distributed throughout the cytoplasm?
5 Does *Paramecium* show any ability to distinguish selectively between particles of nutritional and non-nutritional importance?

INVESTIGATION OF CILIARY ACTIVITY IN THE GILL OF *MYTILUS*

Apparatus
A mussel
Suspension of carmine particles in water
Invertebrate saline
Pipette, scalpel and scissors
Microscope slide and microscope

Method
1 Open the mussel by inserting the blade of a scalpel between the two valves of the shell and severing the posterior adductor muscle which holds the valves together. (This muscle is located at the broader and more rounded end of the animal.)
2 Now cut through the smaller anterior adductor muscle at the other end of the shell, and carefully pull the valves apart. The ligaments at the anterior end will also have to be broken and then the visceral mass running between the valves can be cut away.
NB Keep the blade of the scalpel horizontal so as to avoid damaging the delicate mantle and gills which line the inner surface of the two valves.
3 Using one of the valves, balance it so that it is placed horizontally on a flat surface.
4 Pipette 1 to 2 drops of a suspension of carmine particles onto the middle of the gill.
NB The surface of the gill should be moist in a freshly opened specimen, but if not, add a few drops of invertebrate saline before using the carmine suspension.
5 Note the directions in which the carmine particles move and indicate this on a diagram by means of arrows.
NB Do not add so much of the carmine dye that the mantle cavity is flooded, because it will then be difficult to follow the direction of movement of the particles.
6 Now carefully cut off a small piece of gill, not more than 1 cm wide and supporting the delicate tissue on the blade of a scalpel, place it on a microscope slide.
7 View under the microscope at low power and observe the gill filaments running perpendicular to the long axis of the gill.
8 Add 1 to 2 drops of the carmine suspension and note the direction of flow of the carmine particles, again recording these observations on a diagram of the gill filaments.
9 Using the high power objective of the microscope, view one gill filament and look for the presence of cilia.

Questions
1 How can the results be explained in terms of inhalant, exhalant, feeding and respiratory currents?
2 Is there any noticeable difference in the way in which the cilia deal with large and small particles and if so, what is the functional significance of this?

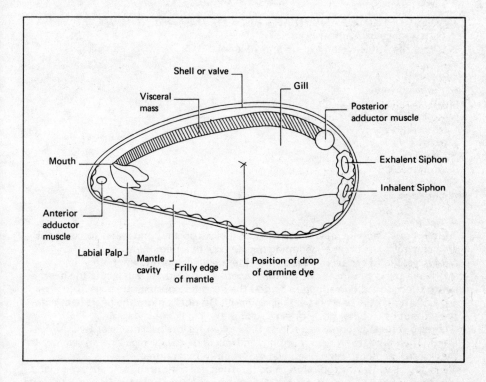

INVESTIGATION OF THE HEART BEAT OF *DAPHNIA*, THE WATER FLEA
A Over a range of temperatures
B Under the influence of Aspirin and Alcohol

Apparatus
Daphnia culture
Aspirin solution (1 Aspirin tablet dissolved in 250 ml of water)
1% and 10% Alcohol solution
Cavity slides
Pipettes
Stopwatch
Thermometer
Filter paper
Microscope

A Method
1 Mount one *Daphnia* on a cavity slide, in just sufficient pond water to cover the animal, whilst limiting its swimming movements to a minimum.
NB Excess liquid can be removed with filter paper or cotton wool.
2 Observe the organism under the microscope at low power and locate the heart which is positoned dorsally, just behind the head (see diagram on page 30).
NB Ensure that the heart is beating regularly. Do not confuse the heart beat with the pulsations of other body organs, such as the trunk appendages.
3 Using a stopwatch, now record the time taken for the heart to beat twenty-five times. Then take two more recordings and calculate the average time in seconds for these three readings, using pond water at room temperature.
NB If the heart beat is too rapid to count, then use a more suitable specimen, and if the three readings fluctuate greatly, then allow some time for the *Daphnia* to settle down.
4 Repeat, using pond water at temperatures of 25, 35 and 45°C (allowing time between each set of readings for the *Daphnia* to settle down again), removing each lot of water by soaking it up with filter paper.
5 Tabulate the results and plot them as a graph, with time in seconds, for twenty-five heart beats along the vertical axis (ordinate) and temperature in degrees centigrade along the horizontal axis (abscissa).

B Method
1 Mount a fresh *Daphnia* in a drop of pond water on a cavity slide and time twenty-five contractions of the heart, taking three readings and averaging them.
2 Using the same specimen, drain off the pond water, replace it by a drop of aspirin solution and record the heart beat as previously.
3 Repeat, using aspirin solutions which have been diluted to half and then a quarter of the original strength (allow time for the specimen to settle down in between each set of readings).
4 Now take a new *Daphnia*, record the rate of heart beat in pond water, and repeat as before, but this time add one drop of 1% alcohol solution.

5 Drain off the 1% alcohol solution and replace it by a drop of 10% alcohol solution, recording the rate of heart beat again and making any relevant observations.

Questions
1 What type of graph is produced when the data from Experiment A is plotted and what conclusions can be made about the influence of temperature on the rate of heart beat?
2 What would you expect to happen at 60°C and above?
3 What effect does a range of concentrations of aspirin solution have on the rate of heart beat?
4 What effect does a range of concentrations of alcohol solution have on the rate of heart beat?
5 Can any comparison be made between aspirin and alcohol in terms of their influence on the heart of *Daphnia*?
6 In Experiment B must a control be included, or is it already present?

INVESTIGATION OF LOCUST HEART BEAT OVER A RANGE OF TEMPERATURES

Apparatus
Locust
Invertebrate saline
Pins, pipettes
1 litre beaker
Plastic petri dish
Stopwatch
Thermometer
Filter paper
Bunsen burner, tripod, wire gauze, asbestos mat

Method
1 Remove the head and thorax of a lightly chloroformed locust. Open up the body cavity by a ventral incision and cut away the gut, reproductive organs and yellow fat. This will expose the dorsally situated heart running down the length of the abdomen.
2 Place the strip of abdominal body wall in a plastic petri dish lined with a layer of moist filter paper and secure the preparation by means of a pin at each end.
NB If it does not lie flat, cut away part of the lateral portions of the body wall, being careful to avoid stretching or pulling, as this may damage the heart.
3 Float the petri dish in a litre beaker filled with water at room temperature and measure the temperature of the water in this water bath.
4 Ensure that the heart is beating regularly and keep it moist with invertebrate saline.
NB Do not cover the heart with any more than a film of liquid in case it becomes flooded.
5 Using the stopwatch, record the time taken for the heart to beat twenty-five times; take two more recordings and calculate the average time in seconds for these three readings taken at room temperature.
6 Repeat this procedure, heating up the water bath to a temperature of 25, 35 and then 45°C. (Allow time between each set of readings for the heart beat to adjust.)
NB Cover the mouth of the beaker whilst the temperature of the water is being raised, so that the air above the preparation is warmed up more rapidly; remove the lid when ready to record the heart beat.
7 Tabulate the results and plot them as a graph with time in seconds for twenty-five heart beats along the vertical axis (ordinate) and temperature in degrees centigrade along the horizontal axis (abscissa).

Questions
1 What type of graph is produced when the data from the experiment is plotted, and what conclusions can be made about the influence of temperature on the rate of heart beat?
2 Do you expect the conclusions to be similar to those for the previous experiment?

INVESTIGATION OF ENZYME ACTIVITY IN THE GUT OF A LOCUST

Apparatus
2 Locusts
1% Starch solution
Benedict's solution
Iodine solution
Invertebrate Saline and distilled water
2 watch glasses and pipettes
Visking tubing
Bulldog clip
Boiling tube, a test tube and beaker
Bunsen burner, tripod, asbestos mat and wire gauze

Method
1 Open up the abdomen of a freshly killed locust, remove the gut and macerate in 6 cm³ of 1% starch solution in a watch glass.
NB Ensure that the locusts have been fed on some plant material such as watercress, before carrying out the experiment.
2 Tie a firm knot in one end of a 6 cm strip of moistened Visking tubing and pipette the entire contents of the watch glass into this tubing, sealing the open end with a bulldog clip.
NB If any liquid has spilt onto the outside of the Visking tubing, wash it off with distilled water.
3 Place the Visking tubing into a boiling tube containing just sufficient distilled water to cover it. Label this A and leave for 30 minutes. *NB* Avoid using excess water.
4 Repeat this procedure, using the gut of a second locust, but first place the gut in a test tube, add a few cm³ of invertebrate saline and put the test tube into a water bath containing boiling water for ten minutes.
5 Now remove the gut from the invertebrate saline, macerate in 6 cm³ of starch solution, place inside the Visking tubing and leave immersed in water in a boiling tube, labelled B, for 30 minutes as before.
6 After 30 minutes remove a few drops of liquid with a pipette from the boiling tubes A and B and test for the presence of starch and reducing sugar. Repeat these tests, using a sample of fluid from inside the Visking tubing. *NB* (i) To test for starch, use a few drops of iodine solution which will turn blue-black in colour if starch is present. (ii) To test for reducing sugars, add 1 cm³ of Benedict's reagent to 1 to 2 cm³ of the test solution and place in a beaker of boiling water for about 5 minutes. A green or yellow or orange colour and precipitate indicates the presence of reducing sugars. (If an expected positive result does not appear, simply increase the quantity of test solution and repeat the test.)

Questions
1 Was starch present in A and B after 30 minutes?
2 Was reducing sugar present in A and B after 30 minutes?
3 What is the significance of the different results for A and B?
4 What function does the Visking tubing perform in this experiment?
5 Is a control necessary in this experiment or is it already present?

INVESTIGATION OF THE POWERS OF REGENERATION OF PLANARIA (FLATWORMS)

Apparatus
Planaria
Raw liver or sausage meat and blood
Pond water
2 petri dishes and lids

Method
1 Sterilise glass petri dishes and dissecting instruments by passing them through the flame of a bunsen burner. (Plastic petri dishes can also be used if they are taken from a sterile unopened packet.)
2 Place some pond water, or even the liquid medium in which the planaria were sent by the suppliers, in each of the three petri dishes.
3 In the first petri dish place a planarian whose head has been divided into two halves by one median longitudinal cut extending one-third of the way down the body, and another planarian which has been divided completely into two halves by one median longitudinal cut extending from head to tail.
4 In the second petri dish place a planarian which has been cut transversely into two halves and another which has been divided into three transverse sections.
5 In each petri dish place a control planarian, which has not been dissected in any way and then partially cover with a lid to prevent excess evaporation of water from the dish, whilst still allowing for the entry of oxygen.
6 The planaria should be fed at regular periods of every three days or so, either on a diet of fresh finely chopped liver or small pieces of sausage meat soaked in blood.
7 It is advisable to change the water in the petri dishes every three days, to remove the solid bits of food and the threads of mucus that the flatworms secrete.
NB If the bottom of the dishes becomes very slimy and messy, it may be necessary to transfer the animals to a completely new petri dish.
8 Observe over a period of a few weeks the degree of ability of the planaria to regenerate or repair themselves.

Questions
1 Which of the sectioned flatworms showed the highest degree of regeneration?
2 Which of the sectioned flatworms exhibited repair rather than regeneration?
3 Which of the sectioned flatworms failed to regenerate or repair themselves?

INVESTIGATION INTO THE BEHAVIOURAL RESPONSES OF PLANARIA TO THE STIMULUS OF
A Light
B Currents in the water

Apparatus
Several Planaria
Pond water or tap water
White and black paper
Cellulose tape
Petri dishes
Pipette
Lamp

A Method
1 Set up a range of choice chambers, using the petri dishes and black and white paper provided, placing one planarian in the centre of each dish in sufficient pond or tap water to cover the bottom of the container. (See diagrams.)

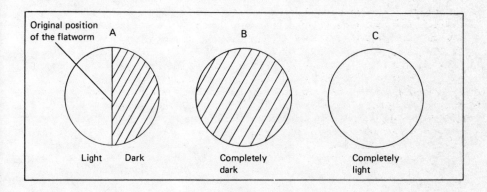

2 Observe the degree of movement exhibited by the flatworms in choice chamber A and whether the movements are directional or random wanderings.
3 Count the number of turns or changes in direction that the planarian makes during a period of time of 2 minutes, in choice chambers B and C.
4 Repeat, using choice chamber C, but this time use a source of bright light above the dish, from a bench lamp.
5 Tabulate the results and plot them as a graph with the number of turns per 2 minutes along the vertical axis (ordinate) and the light intensity, from no light → moderate light → bright light along the horizontal axis (abscissa).

B Method
1 Place one planarian in a petri dish, in pond or tap water, and using a pipette filled with water, gently direct a jet of this liquid at the head, tail and lateral body regions of the flatworm.
2 Observe the response of the flatworm to the stimulus of a current of water and the degree of sensitivity of different parts of the body to this stimulus.

Questions
1 Does planaria respond positively or negatively to the stimulus of light?
2 How did the planaria respond when placed on a completely dark or light background, i.e. in the absence of a directional light stimulus? Was there any noticeable difference in the degree of activity of the flatworms?
3 What can be deduced from the graph about the relationship of activity of the flatworm and light intensity? What is the significance of these results?
4 Do planaria respond positively or negatively to the stimulus of a current of water?
5 Which part of the animal is sensitive to this stimulus or is the entire body surface receptive?
6 What is the significance of this response in the life of the animal?

GENETICS EXPERIMENTS WITH *DROSOPHILA MELANOGASTER* (THE FRUIT FLY)

Life cycle
This is typical for an Endoptergote insect, e.g.
Egg → Larva → Pupa → Imago (adult).
NB Refer to page 60.

The egg
is white and oval, possessing two filaments to prevent it from sinking into the substrate on which it was laid.

The larva
or maggot is very active in its search for food and burrows into the substrate, forming a network of channels throughout the medium. This larva moults twice.

The pupa
is produced when the larva ceases to burrow and feed on the medium, but selects a dry place where its larval cuticle hardens and darkens to form the pupal case or puparium.

The imago
emerges from the pupal case and on contact with the air its body soon dries and darkens in colour. Emergence often occurs between 7a.m. and 11a.m., but this is not always true of flies bred in incubators. Three to 8 hours after emergence the flies are sexually mature and capable of mating and 12 hours after emergence the females are ready to start laying eggs. Once the female has received a supply of sperm from the male, she will use these sperm acquired at the first copulation to fertilise all the eggs she produces throughout her life. An older female may lay unfertilised eggs, but these are always sterile. The whole life cycle takes about 26 days for the females and 33 days for the males.

Sexing
Male fruit flies can be distinguished from females because the male is smaller in size and his abdomen is rounded and black at the tip.

Conversely, the abdomen of the female is larger and more oval and bears faint black stripes across it. (These striations do not fuse as in the male, so the abdomen appears light in colour.)
NB If the adults have only recently emerged from the puparium, their bodies may not have darkened, in which case, sexing must be carried out on the basis of relative size and shape of the abdomen.

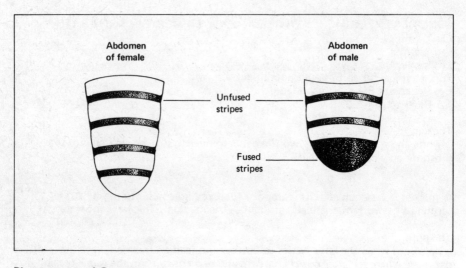

Phenotypes and Genotypes.

Drosophila melanogaster possesses three pairs of autosomal chromosomes and one pair of sex chromosomes, XX in the female and XY in the male.

All the characteristics exhibited by the flies are controlled by genes present on one of the four pairs of chromosomes. The wild type exhibits a standard phenotype and genotype; each of the wild type characteristics being signified by the symbol '+'

The mutant exhibits features which are deviant from the wild type, e.g. differences in size and shape of wings and differences in body and eye colour. The mutant feature is recessive to the wild type feature.

Wing size is an example of autosomal inheritance, because the gene controlling this feature occurs on an autosomal chromosome. The wild type has a standard size wing '+', whilst the mutant has a small vestigial *'vg'*.

Eye colour is an example of sex-linked inheritance, because the genes controlling this feature occur on the sex chromosome (XX or XY).

The wild type has a red eye '+', whilst the mutant has a white eye *'w'*.

INVESTIGATION OF AUTOSOMAL INHERITANCE INVOLVING ONE PAIR OF GENES

Apparatus

Pure-line cultures of normal and vestigial winged *Drosophila*
Culture medium (available ready-prepared from stockist)
Nipagin
Ether
Disinfectant
Flat-bottomed, 3 or 4 inch specimen tubes and perforated plastic stoppers
Plastic dimple tile
Fine paint brush
Cotton wool, filter paper and muslin
Filter funnel and pipette
Lens
Sticky labels
Pressure cooker or autoclave
Incubator (optional)

Method

1 Sterilise the glassware, using the autoclave and disinfect the tile and paint brushes to prevent contamination of the growth.

NB Ensure that no culture medium adheres to the sides of the culture bottles, as they must be absolutely clean and dry. If these points are carefully observed, a culture can last as long as one month.

3 Place a strip of filter paper, 1 cm wide, inside the culture bottle so that the larvae can migrate to a dry region above the medium, at the time of pupation.

NB The filter paper must periodically be replaced, whenever the inside of the bottle becomes damp with condensation.

4 Use a cotton wool plug wrapped in muslin to seal off the opening of the culture bottles. This will allow air to enter, but will prevent the entry of fungal spores (see diagram).

5 The stock flies should, on arrival, be removed from their original culture bottles, sexed, and transferred to new culture bottles labelled stock culture and bearing details of the phenotype and sex.

NB The females have already mated and acquired a store of sperm, so they will continue to produce further pure line offspring. The males may also be used in subsequent experimental crosses.

(a) In order to sex the flies they must be anaesthetised. The anaesthetising chamber, or etheriser, is a flat-bottomed specimen tube fitted with a filter funnel and a plastic stopper. Inside the plastic stopper is a ring of cotton wool on to which two drops of ether have been placed (see diagram).

NB Ether is very volatile and inflammable, so the stopper must be kept on the bottle and the anaesthetisation should be carried out far away from any flame.

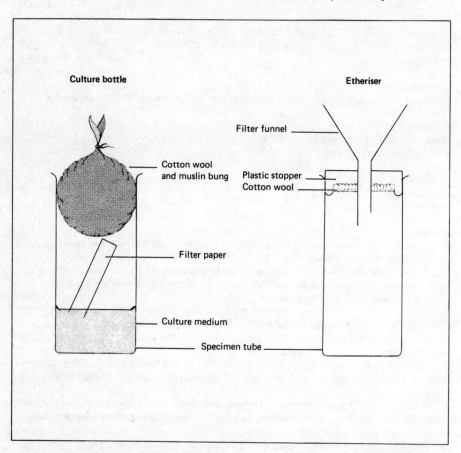

Culture bottle

Etheriser

Filter funnel

Cotton wool and muslin bung

Plastic stopper
Cotton wool

Filter paper

Culture medium

Specimen tube

(b) Now place the funnel back on the specimen tube as shown in the diagram and gently shake all the flies from the original stock bottles into the etheriser.
NB If some flies do not drop readily out of the culture bottles, tap the container against the sides of the funnel and this jolting should dislodge them.
(c) Once all the flies in the etheriser are unconscious they can be shaken out onto a white tile to be sexed.
NB They need only remain in the etheriser for a few minutes and care should be taken to avoid over-anaesthetisation. Dead flies can easily be detected because they assume an unusual posture, the abdomen becoming arched, the wings held stiffly above the back and the legs drawn up under the body.

(d) The flies can now be divided into male and female groups on the white tile and to avoid damage a fine paint brush should be used when moving them about.

(e) Now carefully brush the flies into their bottles, ensuring that they do not fall onto the culture medium, because in an anaesthetised state they will stick in it and be trapped. This can be avoided if the culture bottle is kept horizontal until the flies become active again.

6 Place the culture bottles in an incubator at a temperature of 25°C. If an incubator is not available, the experiment can be carried out in a cupboard, as long as the room temperature does not fluctuate greatly.

NB Above 30°C male flies cease to produce sperm and below 15°C females stop laying eggs.

7 Eggs deposited by the stock culture females will hatch into larvae, pupate and the adults will emerge within a few days of arrival from the supplier. These male and female offspring must be segregated within eight hours of hatching, in order to obtain virgin females which can be used in a cross.

The males and females can then be placed back in their separate culture bottles labelled parental types. (If the females lay no eggs within a few days, then they must be virgins.)

8 The first cross can now be made to obtain the F_1 (filial) generation, by placing three to six normal winged males (++) and three to six vestigial winged females (*vg vg*) into a new culture bottle, which should be labelled parental types and bear details of the cross. A reciprocal cross should also be set up using thee normal winged females (++) and three vestigial winged males (*vg vg*).

NB The number of parent flies used for each cross and the number of culture bottles set up depends on the number of flies available, and obviously, the more crosses that are made, the more reliable the results will be.

9 Once the parent flies have mated and the females have laid eggs, the adults must be removed to prevent breeding between generations. (They can be used to produce more F_1 types.)

NB The parents should be removed within eight days of a cross being set up.

10 When the F_1 generation of adults appear, they must be sexed and the numbers of male and female wild type and vestigial winged types counted.

11 New crosses can now be set up in fresh culture bottles labelled F_1 parental types. The procedure is repeated to obtain the F_2 generation of offspring and again the number of male and female wild type and vestigial winged types must be counted.

Questions

1 How many male and female flies exhibiting a phenotype for standard wing were produced in the F_1 generation?

2 What is the genotype of these flies?

3 Are vestigial winged types likely to occur in the F_1 generation?

4 How many male and female flies exhibiting a phenotype for standard and vestigial wing were produced in the F_2 generation?

5 What are the genotypes of these flies?

6 What is the ratio of standard wing to vestigial wing?

7 Is there any difference between the sexes in the inheritance of this characteristic?

8 What is a reciprocal cross?

INVESTIGATION OF SEX-LINKED INHERITANCE INVOLVING ONE PAIR OF GENES

Apparatus
As before, plus pure line cultures of red- and white-eyed *Drosophila*

Method
1 Separate the stock cultures on arrival into male and female red- or white-eyed types and place in new culture bottles labelled stock cultures.

2 From these cultures collect virgin white- and red-eyed females, when the eggs produced as a result of matings in the original stock culture bottles complete their development to the adult fly.

3 Set up crosses in fresh culture bottles labelled parental types, using three to six red-eyed males $\frac{X+}{Y}$ and three to six white-eyed females $\frac{Xw}{Xw}$

4 Also set up a reciprocal cross between white-eyed males $\frac{Xw}{Y}$ and red-eyed females $\frac{X+}{X+}$.

5 When the F_1 generation of flies hatch out, count the numbers of males and females exhibiting a phenotype for red or white eyes in the two crosses set up.

6 Now proceed with new crosses using F_1 generation flies to obtain the F_2 generation. Again place three to six males and three to six females in new culture bottles labelled F_1 parental types.

7 Collect the F_2 generation and count the number of male and female flies exhibiting red or white eyes.

Questions
1 How many male and female flies exhibiting the phenotype for red or white eyes were produced in the F_1 generation from the two crosses?

2 What are the genotypes of these flies?

3 How many male and female flies exhibiting the phenotype for red or white eyes were produced in the F_2 generation from the two crosses?

4 What are the genotypes of these flies?

5 What are the ratios of red-eyed males and females and white-eyed males and females in the F_1 and F_2 generations?

6 Is there any difference between the sexes in the inheritance of these characteristics?

Glossary

ABORAL SURFACE — opposite surface to that bearing mouth.

ACELLULAR — organism not composed of cells.

ACOELOMATE — three layered (triploblastic) animal with the mesoderm not separated by a coelomic space.

AMITOSIS — cleavage of nucleus into two unequal portions—not involving mitosis during cell division or binary fission.

ANNULI — external rings.

APPENDAGE — extension of the body.

APODOUS — without legs.

AQUATIC — living in water.

ASEXUAL REPRODUCTION — reproduction not involving the fusion of nuclear material of gametes.

BILATERAL SYMMETRY — structurally similar both sides of a median plane.

BINARY FISSION — simplest form of asexual reproduction, the nucleus splits into two by mitosis followed by cytoplasmic cleavage, occurs in the Protozoa.

CARNIVOROUS — feeding on animal matter.

CAUDAL — appertaining to tail.

CEPHALOTHORAX — head and thorax combined.

CERCI — paired appendages on the terminal abdominal segment of Arthropods.

CEREBRAL GANGLIA — aggregations of nerve cells in the anterior region.

CHITIN — horny protective substance as in exoskeleton of Arthropods and perisarc of *Obelia*.

CLOSED BLOOD SYSTEM — blood is usually confined to vessels, seldom allowed to circulate in spaces.

COCOON — protective covering for the eggs of spiders and earthworms, and insect pupa.

COELOMATE — Triploblastic animal with space or coelom present in the mesoderm separating it into outer somatic mesoderm applied to the body wall, and inner splanchnic mesoderm applied to the gut wall. The coelom connects with the exterior via the gonoducts and excretory ducts (coelomoducts).

COELOMODUCT — duct which connects the coelom with the exterior.

COMMISSURE — band of tissue or a vessel linking two structures together.

DIPLOBLASTIC — animal with two germ layers in the embryo—the endoderm and ectoderm—separated by a jelly like substance, mesogloea.

DIRECT DEVELOPMENT — when the animal emerges from the egg as an immature adult, (nymph), there is no larval stage.

ECDYSIS — moulting when the cuticle is shed periodically to allow for growth as in Arthropods.

ENDOSKELETON — internal skeletal tissue.

ENTERON — develops as the second cavity (archenteron), in the embryo and forms the digestive cavity. (The first cavity to develop is the blastocoel)

ENTEROCOEL — coelom formed as an extension of the archenteron enclosed in a mesodermal pouch.

EPIDERMIS — outermost layer or layers of cells formed from ectoderm.

ESTUARINE — aquatic, living in estuaries, able to withstand variations in salinity.

EXOSKELETON — skeletal tissue covering the animal, as in Arthropods.

Glossary

FERTILISATION	— fusion of the nuclear material of two cells or gametes during sexual reproduction, to form a zygote.
FLAME CELL	— flask shaped cells, bearing long flagella, with an intracellular duct, concerned with excretion and osmoregulation in Platyhelminthes.
GASTRIC POUCH	— portion of the enteron delimited by endodermal mesenteries in the Coelenterates.
GONODUCT	— duct from the reproductive organs to the exterior.
HAEMOCOEL	— portions of the primary body cavity (the blastocoel of the embryo) which become filled with blood as in the Arthropods and Molluscs. The haemocoel does not communicate with the exterior.
HERBIVOROUS	— feeding on plant matter.
HERMAPHRODITE	— male and female sexual reproductive organs present in one individual.
HOST	— an individual suffering loss or damage from an association with another individual, the parasite, which may be plant or animal.
HYDROID	— coelenterate, sedentary individual or polyp. Colonies of hydroids may be specialised for different functions.
LARVA	— free living embryo concerned with dispersal, which changes to the adult by metamorphosis.
MARINE	— living in the sea.
MEDUSOID	— free swimming, sexually reproducing individual in the Coelenterates.
MESENTERY	— vertical endodermal partitions as in the enteron of some Coelenterates.
MESOGLOEA	— jelly-like layer cementing ectoderm to endoderm in Coelenterates.
METAMERIC SEGMENTATION (METAMERISM)	— serial division of the animal into segments, each having the same basic structure and origin, but some modified for special functions.
METAMORPHOSIS	— change of form during the life history, may be complete metamorphosis when larva, pupa and adult are present, or incomplete metamorphosis when only nymph and adult are present.
MITOSIS	— division of nucleus followed by cleavage of cytoplasm into two identical portions during cell division or binary fission.
NEPHRIDIA	— primitive, ciliated, tubular, excretory organs, which penetrate the coelom and open to the exterior via nephridiopores, as in Annelids.
NYMPH	— immature individual which emerges from the egg and gradually changes into the adult by incomplete metamorphosis, as in some Arthropods.
ORAL SURFACE	— surface bearing the mouth.
ORGAN	— collection of cells and tissues specialised for a particular function.
ORGANELLE	— portion of cytoplasm specialised for a particular function as in Protozoa.
OPEN BLOOD SYSTEM	— blood not confined to vessels, but circulates in spaces or haemocoels derived from the blastocoel of the embryo, as in Arthropods and Molluscs.
OVIPAROUS	— female lays eggs, the young are not born alive.

Glossary

PARASITE	— individual living in association with another—the host, which suffers some loss or damage. Ectoparasites live on the external surface of the host, endoparasites live within the host.
PENTAMEROUS RADIAL SYMMETRY	— symmetrical about five radial planes.
POLYEMBRYONY	— formation of numerous embryos from a single fertilised egg.
POLYPODOUS	— bearing numerous legs.
PREDACIOUS	— capturing live animals for food.
PREHENSILE	— ability to grasp.
PREY	— animal captured for food by a predator.
PROTANDROUS	— hermaphrodite animal in which the male gonad is mature before the female gonad.
PSEUDOCOELOMATE	— animal with a space between gut and body wall, formed by vacuolation of large mesodermal cells. The vacuoles converge and become fluid filled to form a perivisceral cavity.
PUPA	— stage present during complete metamorphosis, between larval and adult stage in some insects, may be quiescent or active.
RADIAL SYMMETRY	— symmetrial about any radial plane.
SAPROPHYTIC	— feeding on dead, decaying matter.
SCHIZOCOEL	— coelom which develops in the embryo as two dorsolateral splits in the mesoderm which extend ventrally and fuse to form a continuous space, dividing the mesoderm into outer somatic, and inner splanchnic.
SEDENTARY	— fixed, stationary, non motile individual.
SEGMENT	— portion of an animal resulting from metamerism, dividing it up into units, each with the same basic structure, but some segments may be modified for special functions.
SEPTUM	— internal partition separating tissues or cavities.
SESSILE	— non-stalked, attached directly.
SEXUAL REPRODUCTION	— involves the fusion of nuclear material of two cells or gametes at fertilisation.
SYMBIOTIC	— association between two organisms for their mutual benefit.
SYNCITIAL	— cells which have lost their separating cell membranes, so the cytoplasm is continuous with scattered nuclei.
TERRESTRIAL	— living on land, breathing air.
TORSION	— twisting of organs during development as in Gastropods.
TRIPLOBLASTIC	— animal with three germ layers in the embryo, the ectoderm, mesoderm and endoderm.
UBIQUITOUS	— found in all environments, all over the world.
VISCERA	— soft internal organs.
VIVIPAROUS	— young emerge from the female as live individuals, eggs are not laid.